Uhlig · Discovering Polyurethanes

Konrad Uhlig

Discovering Polyurethanes

With 124 Illustrations and 33 Tables

Hanser Publishers, Munich

Hanser/Gardner Publications, Inc., Cincinnati

The Author:
Dr. Konrad Uhlig, Ottweilerstr. 9, D-51375 Leverkusen, Germany

Translated from German by
Amanda J. Conrad
Original title:
Polyurethan Taschenbuch

Distributed in the USA and in Canada by
Hanser/Gardner Publications, Inc.
6915 Valley Avenue, Cincinnati, Ohio 45244-3029, USA
Fax: (513) 527-8950
Phone: (513) 527-8977 or 1-800-950-8977
Internet: http://www.hansergardner.com

Distributed in all other countries by
Carl Hanser Verlag
Postfach 86 04 20, 81631 München, Germany
Fax: +49 (89) 98 12 64

The use of general descriptive names, trademarks, etc., in this publication, even if the former are not especially identified, is not to be taken as a sign that such names, as understood by the Trade Marks and Merchandise Marks Act, may accordingly be used freely by anyone.

While the advice and information in this book are believed to be true and accurate at the date of going to press, neither the authors nor the editors nor the publisher can accept any legal responsibility for any errors or omissions thay may be made. The publisher makes no warranty, express or implied, with respect to the material contained herein.

Die Deutsche Bibliothek – CIP-Einheitsaufnahme
Uhlig, Konrad;
Discovering Polyurethanes: with 33 tables / Konrad Uhlig. – Munich: Hanser, 1998
    Dt. Ausg. u.d.T.: Uhlig, Konrad: Polyurethan-Taschenbuch
    ISBN 3-446-21022-9

1-56990-272-0 Hanser/Gardner Publications, Inc.
3-446-21022-9 Carl Hanser Verlag

©1999 Carl Hanser Verlag Munich
Typeset in the UK by the Alden Group, Oxford
Printed and bound in Germany by Kösel, Kempten

Preface

Between the lightest (145 g) and heaviest (1.56 kg) book on polyurethanes there is a niche for this text. I willingly accepted Carl Hanser Verlag's proposal to follow up "Bayer Polyurethanes" which was published 18 years ago. In contrast to traditional polyurethane publications, however, I chose quite a different format. To spare non-chemists chemical formulae, I begin with Neanderthal Man's axe. It only takes a few lines to cover the intervening 100 000 years up to the present. After a few remarks on the history, market and resources for polyurethanes, we turn to the forms and applications of polyurethanes. After further chapters on their properties and industrial production, we devote ourselves to the question of whether "the chemistry is right". The next two chapters deal with the topical subjects of "quality assurance" and "recycling". They are followed by references to sources of polyurethane expertise. It is inherent in the nature of polyurethanes that some terms will emerge in the "pre-chemical" chapters, but will be explained at a later stage.

This book is intended mainly for existing and potential polyurethane processors who supply diverse markets with products made from raw materials from the chemical industry. This book will not replace technical discussion but will serve as an introduction. The text also contains useful information for end consumers or, more appropriately, end users. The book provides notes about PU raw materials and processing for plant and machinery builders.

This book is also intended for judges, advisors and decision makers, and provides at least an overview of polyurethanes. These include colleagues in organizations, industries, recycling firms, trade unions, parties, consumer groups, etc. *Discovering Polyurethanes* is obviously also addressed to PU-insiders, represented by sales personnel or engineers, who will use it outside their special area to obtain brief information about the overall range of polyurethanes. Finally, it will help PU novices to approach this fascinating material.

Special thanks go to the Polyurethanes Business Group at Bayer AG and to the Research and Marketing Departments for their kind assistance in the preparation of the book. For their assistance with illustrations, proof reading, computer graphics, etc., I would like to

thank my former colleagues and comrades in arms in PU Application
engineering, PU Research, PU Staff and in the Plastics, Paints and
Special Business Groups and in Advertising and Public Relation. My
thanks also go to BASF/Elastogran, Bayer Fasern GmbH, Dow-
Deutschland, ICI-Deutschland, Deutsche Shell, Rhein-Chemie, Elf
Atochem, Hennecke, Krauss-Maffei and the VDK in VCI for the
kind loan of original texts and illustrations. I would like to thank the
publishers, Hanser, Hüthig, Moderne Industrie, Crain Communications
and ICIS-LOR, London, for the licence to use illustrations from their
publications.

Dr. Glenz from Carl Hanser Verlag increasingly appreciated my point
of view and perception of things; Mrs. Sabine Klingan, M.A., polished
some infelicities of style. Grateful thanks are due to them both.

I am also thankful to Mrs Amanda J. Conrad for her outstanding
translation of this technically oriented work, especially considering my
individual (and unusual) expressions and writing style.

Leverkusen   *Konrad Uhlig*

Contents

1 The polyurethane age

Man does not need polyurethanes, knives and forks, shoes and refrigerators, let alone cars, to satisfy his basic needs. He can go hunting and gathering on foot and keep warm in an animal skin. The flesh will keep cool in a cave and can be chopped up with an axe. "Quality of life" was unheard of in the primeval world of our forefathers, thousands of years ago.

Fig. 1. *Homo sapiens neanderthalensis* dressed in a skin, with axe and prey

With progress from the "existence" of the nomadic hunter and gatherer to the "awareness" of settled folk, man has developed a need for convenience and comfort. Potash and sodium sulphide have enabled him to transform a stinking skin-clad board into a soft, warm bed. Nowadays, any nostalgic delight in a bed of straw ends abruptly when bites and stings from unwanted guests complement the prickly stalks. Renewable raw materials from geese and lambs simplified matters considerably. Botanical remedies, too, can be used alongside modern aids: for example, an electronically controlled bed equipped with a

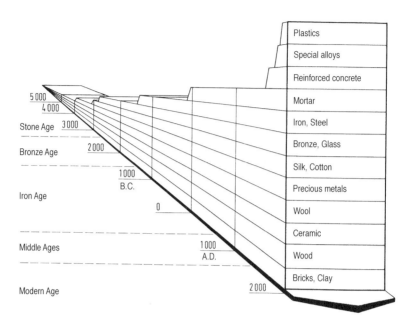

Fig. 2. Materials as indicators of cultural development

panic grass and spelt filled mattresses can apparently soothe asthma [1]. The human imagination can dream up countless designs and uses for materials. Fig. 2 shows various materials as indicators of cultural development, the most recent being the plastics which we love to hate nowadays and which include polyurethanes. Without them, sport and games would only be half as much fun, food would go off, everyday life and car journeys would use up resources, etc.

Polyurethanes have two "drawbacks". First their name, and then the fact that they are usually "invisible". The name "urethane" is an invented word, derived from *ur* ina (Latin urine) and *éthan* ol (French ethanol), and is associated with the French chemist *C. A. Wurtz* (Fig. 3) who first wrote about the preparation of isocyanates in 1848 [2].

"Urethane" is a chemical compound composed of urea and ethanol. This toxic and potentially carcinogenic substance (MAC 1995: III/A2), unwittingly consumed in ng-amounts by lovers of whiskey and yoghurt [3], has given its name to a whole class of chemical compounds (IUPAC

Fig. 3. Charles-Adolphe Wurtz (1817-1884): Discoverer of isocyanates

Rule C-431). It is characterized by the atomic linkage

$$-N-C-O-$$
$$|||$$
$$HO$$

and is also known as an "ester of carbamic acid". If the right-hand "O" is omitted, the structural element of useful products such as silk or wool (proteins) is obtained. Urethanes are formed by attaching other atoms or groups of atoms to this "O" and to the left-hand "N", for example in baking powder (commercial ammonium carbonate), medicines and plant protectives. Mattresses, soles of shoes, swimming trunks, car seats, refrigerators, district heating pipes, etc. consist of polyurethane. They contain many (hence "poly") urethane groups. Strictly speaking, polyurethane should be abbreviated to "PUR" in compliance with official German and International standards (DIN and ISO standards). However, the abbreviation "PU" is more common in English, and even German, texts. We will stick to the normal English version in this pocketbook.

Although polyurethanes were discovered 60 years ago, in 1937, they were concealed (for example in recesses, behind coverings, films, etc.) until the 1970s. This "drawback" has been overcome, and we are now surrounded by "visible" mouldings in window frames, stereo, TV or computer casings, sports equipment, bicycle saddles, etc. Whether "perceptible", concealed or visible, whether an all-PU product, for example an all-foam mattress, or part of the end product: polyurethanes will continue to serve man and his environment. Examples from the fields of food, sports and transport will illustrate this:

- From the North Pole to the Equator, refrigerators are an indispensable part of contemporary life. With its negligible weight, PU rigid foam helps to keep in the cold, prevent food decay and avoid extravagant, environmentally unacceptable wastage of energy.

- Modern skis are based on the wooden runners used by neolithic deer hunters 5000 years ago. Aspiring medallists and aged cross-country skiers now glide smoothly over the snow using equipment with a polyurethane core.

- Whereas Tutankhamen's chariot was produced from renewable raw materials 3400 years ago, because there was no alternative, and the Romans covered their wooden, spoked wheels with iron 1800 years later, fine-grained or PU-grained, renewable raw materials are now a status symbol in limousines, a further 1800 years on. Like normal cars, they now contain 20 kg of polyurethane on average. Of this, a few 100 g support the entire body on the suspension and chassis throughout the life of the car. These obscure, minute parts, made of cellular PU elastomer, are found on the shock absorbers.

2   The PU market and resources

The polyurethane market has evolved on account of the more or less conscious popularity of polyurethanes among consumers. Global strategists are also aware of supply, demand and competition, which can change dramatically from one day to the next, as consumer behaviour is unpredictable. A glance at Fig. 4 shows that this is not surprising when observed over lengthy periods of time. The growth and decline, for example, of the gross national product (GNP) or of a class of products such as polyurethanes is a recurring phenomenon.

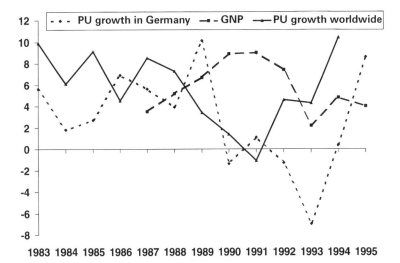

Fig. 4. Economic growth in polyurethanes in the world and in Germany compared with gross national product [%]

In contrast to other plastics markets, for example in PE, PP or PVC, polyurethanes are rarely sold directly but, rather, in the form of raw materials. As we will see in Chapters 4 and 5, the chemical industry supplies liquid chemicals to firms who will foam, inject, spread, pour, mould and spin an extensive range of polyurethanes from them. This is why there is no single market price for polyurethanes. Whereas the raw materials cost about DM 3.00 for 1 kg of flexible foam and twice as much for rigid structural (integral skin) foam, they cost five times as

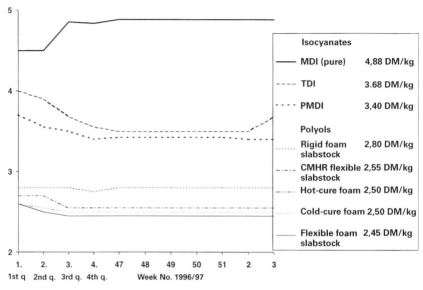

Information based on weekly price report for PU raw materials compiled by
ICIS-LOR Ltd., London, Tel. +44(0) 181 652 3535; Fax +44(0) 181 652 3929

Fig. 5. Prices of PU raw materials (15.1.1997)

much for high-tech top-quality elastomers. The average price of the
main raw materials (Fig. 5) is updated and published every two months
[4], but can be obtained directly from ICIS-LOR at any time.

So-called system suppliers often intervene between raw material
manufacturers and polyurethane manufacturers, hereafter called
"processors" for short. The system suppliers obtain large quantities
of polyurethane raw materials from the chemical industry, for example
in road tankers, and "formulate" ready-to-use systems for small
containers or drums. Depending on their marketing strategy, large
raw material manufacturers may run their own research and development departments, and market their expertise in formulations. The
liquid chemicals are chemically reacted by the PU processor, to form
polyurethanes. The components can be mixed by hand (Fig. 6), and this
is still the normal laboratory method, assisted by electric stirrers, for
test foaming. Polyurethanes can be manufactured industrially using
complete electronically controlled plant (Chapter 5) which is sold by the
machine-building industry, another participant in the polyurethanes
market. Fig. 7 shows the basic structure of the polyurethanes market.

Fig. 6. Otto Bayer (1902–1982) and his team invented the principle of polyurethane production back in 1937. Their manually stirred "foam fungus" is ritually demonstrated at PU presentations

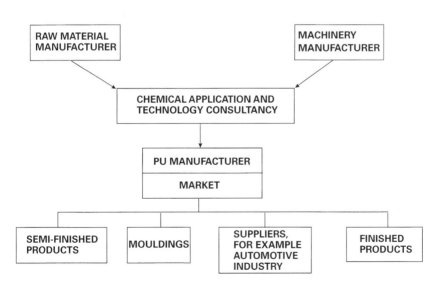

Fig. 7. Structure of the polyurethanes market

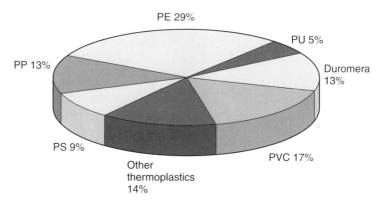

Fig. 8. Global plastics consumption (in %) 1997, total about 142 million tonnes

With a 5 to 6 % share of global plastics consumption, polyurethanes have come fifth for over 20 years (Fig. 8).

Despite past recessions, there has been an immense increase in global PU consumption (Fig. 9).

The main consumers in 1996 were North America with a 25 % share and Europe with a 25 % share, followed by Japan (7 %) and the Far East with 18 % share. The remainder is split between Latin America (7 %), the Middle East and Africa. Fig. 10 shows that almost two-thirds of polyurethanes are used by four main industries: furniture/mattresses, transport (including 0.9 million tonnes in cars), construction and industrial insulation. On the other hand, a mere third is used for "other applications". This is another indication of the wealth of applications for polyurethanes (Chapter 3).

The supply of polyurethane raw materials will be secure; plant capacities will be adapted to expectations. Plant was run to about 80 % of its capacity in 1990. In addition to supply and demand, PU raw material prices are constantly influenced by competition. Understandably, raw materials manufacturers and processors want a larger share of the cake, but this limits their scope with prices and production capacity. The important PU raw material MDI graphically illustrates this (Fig. 11).

If prices rise with demand, capacity is increased to enable more products to be sold at these prices. However, most raw material

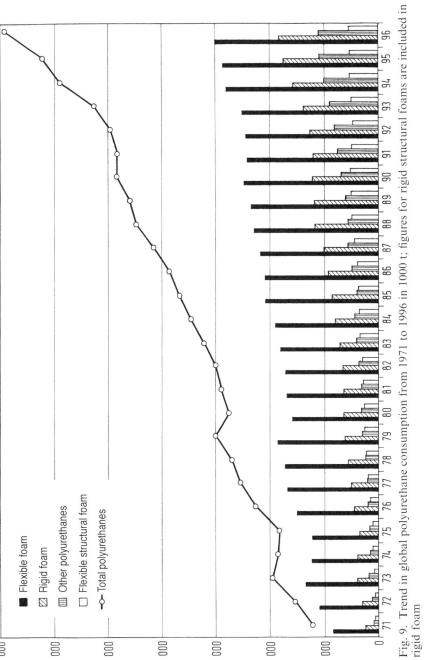

Fig. 9. Trend in global polyurethane consumption from 1971 to 1996 in 1000 t; figures for rigid structural foams are included in rigid foam

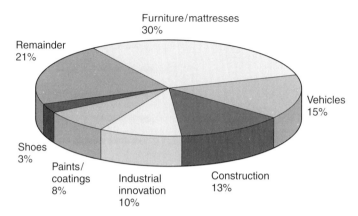

Fig. 10. Global polyurethane consumption in industries (total consumption 6.9 million tonnes/1996)

manufacturers think alike and keep abreast of one another by increasing output. Then they all moan when over-production leads to a drop in prices, and the procedure starts all over again. Joy and despair are inversely proportional among suppliers and customers. An observer would wonder, "What confused economic times we live in" [5]. At the end of the day, it must be remembered that, despite occasional gloom, the future looks rosy for polyurethanes. Chapter 6 gives details about the main protagonists and their raw materials.

We know that polyurethanes are durable products. They do not normally have to be scrapped until the article of which they are part

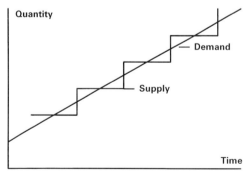

Fig. 11. Interplay between demand and increase in output (applied to the PU raw material MDI)

is scrapped. A refrigerator will break down mechanically before its PU rigid foam insulation crumbles: PU will always be PU! Longevity makes a product environment-friendly and conserves resources. For disposal, polyurethane can be converted in accordance with various regulations: in the form of PU fragments, shreds, flakes or powder or as an energy supplier in proper incinerators or, alternatively, it can be beaten back by the "Chemical mace" into its original components to yield regenerated raw materials. More of this in Chapter 8.

Original PU raw materials (see Section 6.3) are derived from natural resources: rock salt, coal and petroleum, sugar cane or beet, corn, fat – to name but a few. Although the latter are renewable raw materials, we perceive the first three as finite resources. Global supplies of rock salt are estimated at 3700 billion tonnes [6] and of coal at 7000 billion ($=10^{12}$) tonnes [7]. On the other hand, the 136.9 billion tonnes [8] of global crude oil reserves which can be extracted by conventional methods are quite low with an annual extraction (1994) of 3.2 billion tonnes. It is not known whether a new method of extraction will be

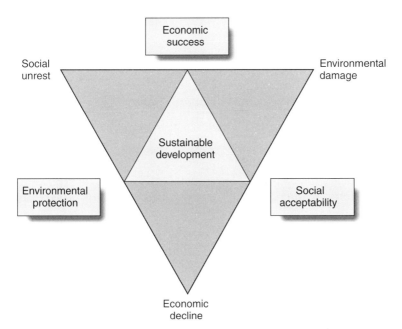

Fig. 12. Sustainable development factors (see also Fig. 108)

available by 2036 for developing the further known 490 billion tonnes of oil (cf. Table 28). This is a good reason to proceed cautiously with the quantities available for the foreseeable future. Civil aviation consumes about 5 % of annual petroleum production [9].

Only 4 % of the extracted petroleum is used for global plastics production (about 130 million tonnes, see Fig. 8). This is roughly 131.5 million tonnes, of which 5 % or 6.6 million tonnes or *0.2 % of the petroleum extracted annually are used for PU!* So we can continue to produce it in an environment-friendly and socially acceptable manner within economic constraints: sustainable development in practice (Fig. 12).

3 Classification of types

Living examples provide more information about the usefulness of a material than numbers, graphs and statistics. Concrete applications strikingly demonstrate the value of a substance from which an article is produced. The extensive but incomplete range of polyurethanes listed in this chapter might arouse enthusiasm, a deep impression, amazement or even disbelief [10]. Polyurethanes are obviously competing with other, so-called "natural", materials. Rather than explaining all the technical and psychological criteria for or against a particular material, we will simply "absorb" the wealth of applications for PU.

3.1 Forms of polyurethane

An impressive array has evolved since the dawn of the PU era. About 80 % of all polyurethanes are used in the form of foams, 20 % in the form of non-cellular products. The various forms of polyurethane are shown in Fig. 13.

Flexible and rigid foams are the best known forms of PU. They are rarely seen except as a colourful stack of mats in department stores or DIY centres. They are normally concealed by textile coverings, plaster-board and plastic panels, sheet metal, foils, wrapping paper and other outer layers. On the other hand, the PU is "visible" in *structural foams* (integral skin foams) where the non-cellular surface layer is produced round the cellular core in a single operation from a single material. *Non-cellular polyurethanes* are also visible, though usually only to engineers. However, the end-user can "feel" some of them in elegant PU-treated leather goods, PU-coated parquet floors or, last but not least, in textiles containing PU fibres. He can experience the range of polyurethanes in his car (Fig. 14).

With 16 to 30 kg per car, they are the plastic most commonly used for seats, head-rests, sound insulation, etc. in the automotive industry. A further 15 to 20 kg of PU is used in the form of paints, pigments, coatings and underseals.

So polyurethanes increase the longevity and reduce the weight of cars

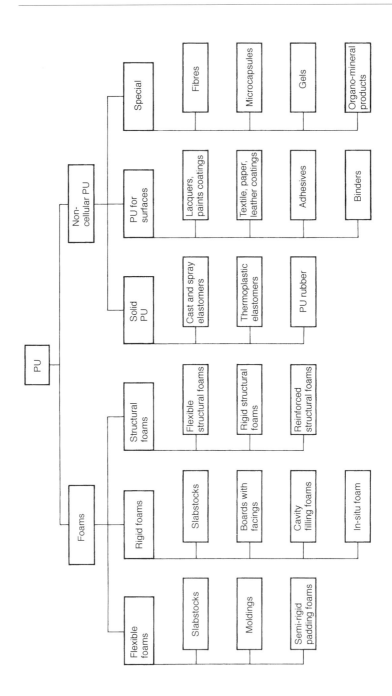

Fig. 13. Forms of polyurethane

○ *PU flexible foam*
◌ *Semi-rigid padding foam*
▲ *Semi-rigid structural foam*

▼ *Rigid structural foam*
∪ *Elastomers*
■ *PU rigid foam*

Fig. 14. Practical examples of PU on or in cars

and therefore save fuel. According to recent estimates, a reduction of 100 kg in weight reduces consumption by 0.2 to 0.5 l/100 km. With a potential reduction in sheet thickness from the present 0.75 mm to 0.65 mm, a total weight reduction of 200 kg, that is about 25 %, is feasible. A prototype is expected in 1998 [11].

3.2 Applications according to form

The concrete examples listed alphabetically according to form in Table 1 illustrate the applications for each polyurethane. The host of applications essentially speaks for itself, so this section merely summarizes the main areas. However, readers should take a look at Fig. 10.

Flexible foam is required by designers and users, depending on the envisaged application. PU comes in many grades: from super-soft for cuddling to rock-hard for ascetics – not hard like a plank, but what the specialists call "a harder quality" of flexible foam.

Flexible slabstocks are used for upholstered furniture (40 %), mattresses (25 %), vehicle trims (20 %), textiles (5 %) and other applications

(10 %), for example in the household, in packaging, etc. According to a major slabstock foam manufacturer, the cushioning in a finished item of furniture costs about 20 %, and the prices of cushioning materials diminish as follows: steel springs > latex > PU slabstock foam. This tendency reflects actual cost factors not yet acknowledged [12]. The scrap created during manufacture of the slabstock is shredded and included as flocking in the spectrum of flexible foam applications (see Chapter 8).

Flexible PU moulded foams are used mainly as seats in road vehicles and aircraft, but also in upholstered furniture and industrial articles. A viscoelastic variant of flexible PU moulded foams provides an excellent material for sound insulation.

Semi-rigid padding foams are almost always encountered in the automotive industry, behind coverings. They combine the main aim of increasing internal (passive) and external (active) safety with an attractive appearance and feel. PU safety cushions are not as "beautiful" but are highly effective in preventing side impacts in the doors of modern cars.

Rigid foams are used in applications where energy losses and therefore CO_2 emissions are to be minimized. For example, about 100 million tonnes of CO_2 per year, that is >10 % of the total output in Germany, could be saved by thermal refurbishment of old buildings [13]. Applications therefore focus on low/high-temperature insulation and lightweight construction: spheres for PU rigid foam are the refrigerated distribution chain from manufacturer to user, the construction and automotive industries, the transport sector, ship-building and packaging. Versatile PU rigid foam is even used in horticulture, in furniture and sports equipment, in solar technology and finally for rock consolidation in mining and civil engineering.

Rigid foam slabstock is a semi-finished product. Insulation panels and half-shells for pipes are cut from it, and parts are also produced for interior vehicle trims (roof liners). *Laminated rigid foam panels* are used for roof insulation and dry plaster boards or as sandwich panels for factory buildings and cold stores, partitions, etc. in the construction industry.

Rigid cavity filling foams are used as high/low-temperature insulating materials in refrigerators and freezers of any type, cold stores and cold-storage cells, for hot water appliances and district heating pipes.

In-situ foam is used for roof and wall insulation, window and door frame fixings, joint seals and as a variant of spray foam for roof sealants.

In contrast to rigid and flexible foams as semi-finished products which can be converted, *structural foams*, sometimes finish as integral skin foams are produced and used almost exclusively as a final moulding. If necessary, they can be given a surface finish (paint, galvanization, etc.). The cellular core contained in the moulding can vary from an optically perceptible cell structure through microcellular structures to zero, in which case the structural *foam* becomes a *solid* moulding. Glass fibre-reinforced variants also have interesting applications: car body elements (for example, mudguards) which can be painted "on-line" at 190 °C by cathodic dip coating can be obtained by thin wall technology (= <2.5 mm wall thickness).

Flexible structural foams are found mainly in the automotive industry, from car steering wheels, through bumpers to bicycle saddles, and in soles of shoes. *Rigid structural foams* are used as engineering polymers for a variety of industrial parts, building, furniture, electrical and sports/leisure articles. Even the comprehensive list in Table 1 is incomplete in this case.

Structural foams graded from cellular to "solid" could only lead to superficially non-cellular PU if there were no significant differences in the process, details of which will be given in Chapter 5. Whereas foams are based on a cell structure which can be deliberately adjusted, a completely homogeneous, i.e. pore-free, matrix is required when using non-cellular polyurethanes. We will see cases where exceptions confirm the rule. Non-cellular polyurethanes are used mainly in industry and are associated with a particularly high added value. Remember, for example, that automobile suspension and life expectancy are considerably improved by a mere few hundred grams of PU elastomer.

Cast and spray elastomers are used in almost all spheres of industry where high performance, low wear and resilience or damping and corrosion protection are critical. All-PU coverings are used in public occupancy applications, for example in the Olympia Stadium in Munich or in the PU cast tyres for cars conceived and tested about 50 years ago but never produced commercially: poor traction on solid road coverings, build-up of heat and the expensive radial reinforcement impeded mass production. But not completely: undaunted engineers attempted to build a car tyre carrying not air but a steel spiral spring embedded in PU elastomer. An inbuilt air cooling system was intended to counteract the build-up of heat [14]. PU casting resins, a variation of solid polyurethanes, filled with ground quartz are used almost exclusively as an insulating embedding compound in electrical engineering.

Thermoplastic PU elastomers (TPU) are designed to meet industrial needs, in the form of injection mouldings, blow-moulded parts, films or coverings and hosepipes, as well as day-to-day requirements in the home, in leisure and sports activities, in medical technology and even in the meadows (earmarks on calves) [15].

PU rubber is a technical variation of PU elastomers which do not require special PU plant but can be processed on existing rubber-manufacturing machinery. The fields of application of PU rubber are comparable, and sometimes identical, to those of cast elastomers and TPUs.

Non-cellular PU for surfaces is applied in thin layers (<1 mm), whether for technical reasons, for example corrosion prevention, and/or for optical reasons. It includes PU adhesives and special polyisocyanates used for producing chipboard.

PU lacquers are used not only for painting wood (furniture, parquet flooring), but also for road and rail vehicles, aircraft and ships, furthermore in engineering for coating metal and in construction for mineral primers.

PU coatings are widely used in textile finishing and coating, bonding of non-woven fabrics, synthetic leather production, paper coating and gluing and for leather dressing.

PU adhesives are used in the shoe and clothing industry, in plastics processing, in the production of packagings and in the automotive and construction industries. PU *binders* are related to PU adhesives but there is a clear distinction. Whereas PU adhesives are used for gluing materials, new materials can be made by using PU binders along with odd-sized particles of solid materials, for example from waste. Examples include chipboard from agricultural and forestry waste (straw, wood shavings) and other articles from industrial waste products (granulated rubber, foam scrap). However, PU binders are also used in metal foundries for foundry resins, as glass fibre sizes and for producing vermiculite panels in the construction industry.

Among the *special PU materials*, *PU fibres* or *elastanes* are well known in textiles.

PU microcapsules are less striking. They are used for producing carbon-free sets of forms or in plant protection as a repository for active ingredients.

PU gels contain up to 90 % of water and are used as odoriphores and moisturizers and also as hot and cold compresses.

Organo-mineral products in the stricter sense include foamed polyurea polysilicates based on waterglass; they are used as rock-consolidating materials in mining.

Table 1. Polyurethane applications according to forms

Applications	Type of PU	Remarks
Adhesion to buildings	Adhesive	Wood, plastic, min. fibres
Adhesive substances	Adhesive	
Automotive interior trims	Adhesive	Aqueous dispersion
Clothing	Adhesive	Flexible foam composite
Flocking	Adhesive	Electrostatic
Liquefied gas tankers	Adhesive	Rigid foam panels/ LNG tanks
Plastics processing	Adhesive	
Shoe soles	Adhesive	
Chipboard form. free	Binder	Of wood, straw, etc.
Foundry cores	Binder	Foundry resins
Glass fibre sizes	Binder	Aqueous medium
Rock consolidation	Binder	Mining, civil engineering
Automotive underseals	Coating	2-pack
Awnings	Coating	
Carrier bags	Coating	Printable sheets
Duplicating paper	Coating	Microcapsules
Leather (tanning, dressing)	Coating	Dyeing, impregnation, finishing

Table 1. Continued

Applications	Type of PU	Remarks
Leather for bags	Coating	
Outer garments	Coating	Dry cleanable
Packaging (paper, card)	Coating	Aqueous dispersions
Patent leather	Coating	"Cold lacquer"
Protective clothing	Coating	
Rainwear	Coating	
Shoe uppers	Coating	For leather, textiles
Split leather	Coating	
Synthetic leather	Coating	Textile backing mats
Synthetic leather, poromeric	Coating	Breathable
Tent roofs, floors	Coating	
Upholstery leather	Coating	
Automotive seals	Elastomers	Hot cast
Automotive spring aids	Elastomers	Hot cast
Bath tubs (acrylic)	Elastomers, reinforced	Cold spraying
Boats and fittings	Elastomers	Cold spraying
Carpet underlays	Elastomers	Cold cast
Conveyor belts	Elastomers	Cold/hot cast or knife application
Corrosion prevention	Elastomers	Cold cast, spraying
Couplings	Elastomers	Hot cast
Crane buffers	Elastomers	Hot cast
Cylinders for ind. use	Elastomers	Hot and cold cast
Deburring drums	Elastomers	Hot cast
Diaphragms, industrial	Elastomers	Hot cast
Drive belts	Elastomers	Hot cast
Energy absorbing parts	Elastomers	Hot/cold cast
Fenders	Elastomers	Hot cast; boats, quaysides
(Fire) hose coatings	Elastomers	Cold knife appln., hot curing
Floors for sporting activities, including foamed floors	Elastomers	Cold cast; halls, outdoors
Grinding discs	Elastomers	Hot and cold cast
Gully floors	Elastomers	Cold spraying
Hydrocyclones	Elastomers	Hot cast
Office furniture	Elastomers	Cold cast
Pipe coatings (inc./ext.)	Elastomers	Cold/hot cast
Pipe scrapers	Elastomers	Hot cast
Playgrounds (children, school)	Elastomers	Cold cast
Polishing discs	Elastomers	Hot cast

▶

Table 1. Continued

Applications	Type of PU	Remarks
Pump linings	Elastomers	Hot cast
Pump stators	Elastomers	Hot cast
Roller blades	Elastomers	Hot cast
Roller-skate wheels	Elastomers	Hot cast
Rollers for ind. Use	Elastomers	Hot/cold cast
Safety glass	Elastomers	Cold cast
Sealant for buildings	Elastomers	Cold spraying
Seals	Elastomers	Hot/cold cast
Shoe inserts	Elastomers	Hot cast
Shower trays (acrylic): reinforced	Elastomers	Cold spraying
Sieves	Elastomers	Hot cast
Skateboard wheels	Elastomers	Hot cast
Sound insulation	Elastomers	Cold cast
Shuttering mats	Elastomers	Cold cast
Spiral pneumatic sifters	Elastomers	Hot cast
Spring elements	Elastomers	Hot cast; solid, foamed
Surfboards	Elastomers	Cold cast
Table surrounds	Elastomers	Cold cast
Toothed belts	Elastomers	Hot cast
Anorak fabrics	Fibres	
Bathing fabrics	Fibres	
Corsetry	Fibres	
Lingerie	Fibres	
Stockings, socks	Fibres	
Textile knitted fabrics	Fibres	
Textile ribbons	Fibres	
Textile woven fabrics	Fibres	
Textile yarns	Fibres	
Trouser fabrics	Fibres	
All-foam furniture	Flexible foam	Slabstock, moulded
Automotive back-rests	Flexible foam	Moulded slabstock
Automotive energy absorption	Flexible foam	Semiflex padding foam
Automotive filters	Flexible foam	Slabstock, moulded
Automotive head-rests	Flexible foam	Moulded
Automotive internal protective padding	Flexible foam	Slabstock, moulded
Automotive seats	Flexible foam	Moulded, slabstock, flocking

▶

Table 1. Continued

Applications	Type of PU	Remarks
Automotive side impact protection	Flexible foam	Semiflex, energy-absorbing
Automotive sound absorption	Flexible foam	Moulded, flocking
Automotive visors	Flexible foam	Slabstock, moulded, padding foam
Balls	Flexible foam	Slabstock, moulded
Bicycle saddles	Flexible foam	Moulded
Carpet underlays	Flexible foam	Slabstock, flocking
Energy-absorbing parts	Flexible foam	Moulded, flocking
Facings	Flexible foam	Slabstock
Filters for industry	Flexible foam	Crosslinked slabstock
Gym mats	Flexible foam	Bonded foam
Hospital equipment	Flexible foam	Slabstock, moulded
Mattresses/cores	Flexible foam	Slabstock, moulded, bonded
Office furniture	Flexible foam	Slabstock, moulded, flocking
Packaging	Flexible foam	Slabstock, moulded, flocking
Padding	Flexible foam	Slabstock, moulded, bonded
Pipes, heat-carrying	Flexible foam	Slabstock
Seals	Flexible foam	Slabstock, moulded
Seating	Flexible foam	Slabstock, moulded
Shoe inserts	Flexible foam	Slabstock
Shoulder cushions	Flexible foam	Slabstock
Sound absorption	Flexible foam	Moulded, flocking, low-exp. foam
Sound insulation	Flexible foam	Slabstock, moulded
Shuttering mats	Flexible foam	Slabstock
Sponges	Flexible foam	Slabstock
Tennis training walls	Flexible foam	Slabstock
Textiles, laminated	Flexible foam	Slabstock, web form
Aircraft	Lacquer	External paints
Automotive interior paints	Lacquer	Aqueous 2-pack
Automotive off-road cars	Lacquer	External hard top
Automotive repair paints	Lacquer	Water-thinnable
Building paints	Lacquer	Concrete, plaster
Corrosion prevention	Lacquer	Surfaces of any type
Electrical insulating paints	Lacquer	
Furniture	Lacquer	

▶

Table 1. Continued

Applications	Type of PU	Remarks
Metallic effect	Lacquer	Automotive industry
Paints	Lacquer	Surfaces of any type
Parquet flooring	Lacquer	Sealing, ready-made parquet flooring
Railways	Lacquer	External paints
Stoving lacquers	Lacquer	
Automotive interior trims	Rigid foam	Thermoformable
Automotive lorry superstructures	Rigid foam	Modular construction
Automotive roof liners	Rigid foam	Thermoformable
Bathrooms	Rigid foam	Plumbing blocks
Boats and fittings	Rigid foam	
Bonding foam	Rigid foam	Also "can foam"
Bricks, hollow	Rigid foam	Integrated heat insulation
Cold storage cells	Rigid foam	
Cold stores	Rigid foam	
Construction of industrial buildings	Rigid foam	PU steel sandwich panels
Containers of any type	Rigid foam	Oil, water, cereals
Containers incl. residential	Rigid foam	Also on-site spraying
District heating pipes	Rigid foam	
Doors, gates, heat insulated	Rigid foam	
External wall insulation	Rigid foam	Heat skin, panels
Facings	Rigid foam	Thermoformable
Flat roofing	Rigid foam	Panels, moulded foam
Flower arranging foam	Rigid foam	Hydrophilic, open-cell
Flower arranging substrate	Rigid foam	
Freezers, see "refrigerated..."	Rigid foam	
Heat insulation for floors	Rigid foam	
Hot-water cylinders	Rigid foam	
Liquefied gas transport	Rigid foam	LNG cold insulation
Multistorey car parks	Rigid foam	Panels, moulded foam
Padding	Rigid foam	Thermoformable
Petroleum tanks	Rigid foam	
Pipelines	Rigid foam	Petroleum etc.
Pipes, cold-carrying	Rigid foam	
Pipes, heat-carrying	Rigid foam	
Pitched roofs	Rigid foam	Panels
Refrigerated appliances	Rigid foam	Refrigerators, freezers
Refrigerated containers	Rigid foam	Also ships
Refrigerated trucks	Rigid foam	Heat insulation

Table 1. Continued

Applications	Type of PU	Remarks
Refurbishment of old buildings	Rigid foam	
Rolling doors, heat insulated	Rigid foam	
Seat frames	Rigid foam	
Seating	Rigid foam	Thermoformable
Shutters/shutter boxes	Rigid foam	Hot/cold insulated
Surfboards	Rigid foam	Hot/cold insulated
Thermal containers for camping	Rigid foam	
Wall heat insulation	Rigid foam	
Water heaters	Rigid foam	
Window seats	Rigid foam	Hot/cold insulated
Automotive air-cond. units for lorries	Rigid structural foam	Casing
Automotive back-rests	Rigid structural foam	Glass mat reinforced
Automotive bumpers rear/front	Rigid structural foam	Glass mat reinforced
Automotive cavity sealants	Rigid structural foam	Glass mat reinforced
Automotive dashboards	Rigid structural foam	Glass mat reinforced
Automotive engine housing	Rigid structural foam	Glass mat reinforced
Automotive interior trimes	Rigid structural foam	
Automotive lorry superstructures	Rigid structural foam	
Automotive roof linears	Rigid structural foam	Glass mat reinforced
Automotive seat shells	Rigid structural foam	Glass mat reinforced
Automotive spare wheel recesses	Rigid structural foam	Glass mat reinforced
Automotive spoilers	Rigid structural foam	Also glass mat reinforced
Automotive tractor cab	Rigid structural foam	
Back-drops	Rigid structural foam	"direkt": Nov. 1996
Boats and fittings	Rigid structural foam	
Cash dispensers	Rigid structural foam	
Cinema seats	Rigid structural foam	
Computer casings	Rigid structural foam	
Coverings	Rigid structural foam	For industrial plant
Dental equipment	Rigid structural foam	
Doors for drinks dispensers	Rigid structural foam	
Facings	Rigid structural foam	
Filter mat carriers	Rigid structural foam	For sewage treatment
Gullys	Rigid structural foam	
Hospital equipment	Rigid structural foam	

▶

Table 1. Continued

Applications	Type of PU	Remarks
Invalid's hoists	Rigid structural foam	
Office furniture	Rigid structural foam	
Petrol pump covers	Rigid structural foam	
Photo-developer accessories	Rigid structural foam	Tanks, rollers
Plant containers	Rigid structural foam	
Plumbing	Rigid structural foam	
Reinforcements	Rigid structural foam	
Seat frames	Rigid structural foam	
Seating	Rigid structural foam	
Shoe trees	Rigid structural foam	
Shop fittings	Rigid structural foam	
Shower cubicles	Rigid structural foam	
Skis	Rigid structural foam	Also water skis
Skittle balls	Rigid structural foam	
Sledges	Rigid structural foam	Including bobsleighs
Stable-venting elements	Rigid structural foam	
Surfboards	Rigid structural foam	
Swimming pool cleaning equipment	Rigid structural foam	
Television casings	Rigid structural foam	
Toys	Rigid structural foam	
Traffic light casings	Rigid structural foam	Euro-Tunnel
Tubbing rings	Rigid structural foam	
Vacuum cleaner casings	Rigid structural foam	
Windows	Rigid structural foam	
Automotive filters	Rubber	
Automotive seals	Rubber	
Automotive spring aids	Rubber	
Bellows	Rubber	
Couplings	Rubber	
Diaphragms, industrial	Rubber	
Energy-absorbing parts	Rubber	
Grinding discs	Rubber	
Moulds	Rubber	
Polishing discs	Rubber	
Pump stators	Rubber	
Rollers for industrial use	Rubber	
Roller-skate wheels	Rubber	
Toothed belts	Rubber	
Automotive arm-rests	Semiflex padding foam	Behind facing
Automotive bumpers rear/ front	Semiflex padding foam	Behind elastic sheath

▶

Table 1. Continued

Applications	Type of PU	Remarks
Automotive cavity sealants	Semiflex padding foam	
Automotive column covers	Semiflex padding foam	Behind facing
Automotive consoles	Semiflex padding foam	Behind facing
Automotive dashboards	Semiflex padding foam	Behind facing
Automotive door linings	Semiflex padding foam	Behind facing
Automotive knee cushions	Semiflex padding foam	Behind facing
Automotive visors	Semiflex padding foam	Behind facing
Energy-absorbing parts	Semiflex padding foam	Behind facing
Automotive aprons	Semiflex padding foam	
Automotive back-rests	Semiflex padding foam	
Automotive bumpers rear/front	Semiflex structural foam	(R)RIM
Automotive disc covers	Semiflex structural foam	RIM
Automotive door leaves	Semiflex structural foam	(R)RIM
Automotive external claddings	Semiflex structural foam	(R)RIM
Automotive ext. protection	Semiflex structural foam	
Automotive external trims	Semiflex structural foam	(R)RIM
Automotive handles	Semiflex structural foam	
Automotive head-rests	Semiflex structural foam	
Automotive interior protective padding	Semiflex structural foam	
Automotive lever knobs	Semiflex structural foam	
Automotive mudguards	Semiflex structural foam	(R)RIM
Automotive spoilers	Semiflex structural foam	Also in (R)RIM
Automotive steering wheels	Semiflex structural foam	Cover
Automotive thresholds	Semiflex structural foam	(R)RIM
Automotive tractor seats	Semiflex structural foam	
Beer barrels	Semiflex structural foam	Cover
Bicycle saddles	Semiflex structural foam	RIM
Facings	Semiflex structural foam	
Fenders for sports boats	Semiflex structural foam	
Grips	Semiflex structural foam	Ski sticks
Hospital equipment	Semiflex structural foam	
Ice hockey boots	Semiflex structural foam	
Ice hockey gloves	Semiflex structural foam	
Inner shoes for skis	Semiflex structural foam	
Office furniture	Semiflex structural foam	
Padding	Semiflex structural foam	For spots equipment
Shoe inserts	Semiflex structural foam	
Shoe soles	Semiflex structural foam	Multi-purpose

Table 1. Continued

Applications	Type of PU	Remarks
Automotive ball sockets	Thermoplastic	
Automotive bearing bushes	Thermoplastic	
Automotive lever knobs	Thermoplastic	
Automotive sealing bellows	Thermoplastic	
Automotive thresholds	Thermoplastic	Also glass fibre reinforced
Cable sheathing	Thermoplastic	For all types of cable
Condoms	Thermoplastic	UT June/July 96, p. 3
Diaphragms, industrial	Thermoplastic	
Earmarks	Thermoplastic	For (grazing) animals
Football boot soles	Thermoplastic	TPU studs
Hosepipes	Thermoplastic	Also for the fire brigade
Ice hockey boots	Thermoplastic	
Rollers for industrial use	Thermoplastic	
Sheets	Thermoplastic	
Sieves	Thermoplastic	Mining, gravel, etc.
Ski boots	Thermoplastic	
Toothed belts	Thermoplastic	

Fig. 15. Cushions made of flexible PU foam

Fig. 16. Car seat cushions made of flexible moulded PU foam

Fig. 17. The dashboard is made totally from PU: support, sprayed skin, padding foam and hot air duct (see Fig. 123)

Fig. 18. Side impact protection made of energy-absorbing PU padding foam PU rigid foam insulation

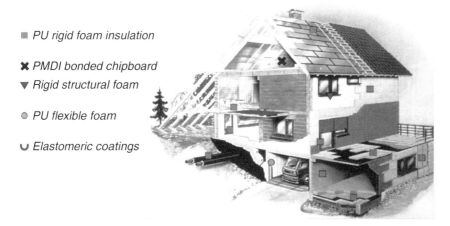

■ *PU rigid foam insulation*

✖ *PMDI bonded chipboard*

▼ *Rigid structural foam*

◉ *PU flexible foam*

↻ *Elastomeric coatings*

Fig. 19. PU in buildings: Energy conservation with PU insulating materials

Fig. 20. The "refrigerated distribution chain": PU rigid foam separates hot from cold

Fig. 21. Pipelines insulated with PU rigid foam transport hot thin petroleum faster than cold thick petroleum

Fig. 22. The yellow PU box on wheels: The German Post Office uses light PU rigid foam in the back of 9000 delivery vans

Fig. 23. Brochures refer to "foamed steering wheels". They are made of semi-flexible PU structural foam

Fig. 24. This prototype of a mudguard made of fibre-reinforced (RRIM-) polyurethane is less sensitive to impacts than steel; it withstands temperatures up to 190 °C and is therefore suitable for cathodic dip coating

Fig. 25. Robust in daily use: coffee machine casing made of rigid PU structural foam

Fig. 26. Tennis shoe soles made of flexible structural foam and running track surfaces made of cold-cure elastomer; the skis have a core made of rigid RIM-PU

Fig. 27. Rigid PU structural foam for backdrops and stage decoration (*G. Collet*, (Büsing & Fasch, Oldenburg, and *D. Jakob*, Bayer AG, are feeling the PU crocodile's teeth from "Prince Ironheart")

Fig. 28. The spring aid in the steel spiral spring consists of high-performance PU elastomer

Fig. 29. PU fun! The roller coaster wheels are made of hot-cast PU elastomer

Fig. 30. Earmark made of thermoplastic polyurethane

Fig. 31. Durable PU corrosion prevention on land and sea vehicles. The hulls of the freighters "Robin Hood" and "Nils Dacke" are protected by a one-pack PU coating. The tram has an environment-friendly paint coating

Fig. 32. Polyurea-bonded chipboard in the form, for example, of floor panels is produced using PMDI and is therefore free from formaldehyde

Fig. 33. The segmented molecular structure and entropy elasticity of PU filaments guarantee a perfect fit

3.3   A to Z of applications

The 294 applications from Table 1 are listed alphabetically in Table 2 so
the appropriate type of PU for a given application can be found.
Creative readers will find suggestions for their own innovations.

Table 2. A to Z of PU applications

Applications	Type of PU	Remarks
Adhesion to buildings	Adhesive	Wood, plastic, min. fibres
Adhesive substances	Adhesive	
Aircraft	Lacquer	External paints
All-foam furniture	Flexible foam	Slabstock, moulded
Anorak fabrics	Fibres	
Automotive air cond. units for lorries	Rigid structural foam	Casing
Automotive aprons	Semiflex struct. foam	
Automotive arm-rests	Semiflex padding foam	
Automotive back-rests	Rigid structural foam	Glass mat reinforced
Automotive back-rests	Semiflex struct. foam	
Automotive back-rests	Flexible foam	Moulded slabstock
Automotive ball sockets	Thermoplastic	
Automotive bearing bushes	Thermoplastic	
Automotive bumpers	Rigid structural foam	Glass mat reinforced
Automotive bumpers rear/front	Semiflex struct. foam	R(RIM)
Automotive bumpers rear/front	Semiflex padding foam	Behind elastic sheath
Automotive cavity sealants	Rigid structural foam	Glass mat reinforced
Automotive cavity sealants	Semiflex padding foam	
Automotive column covers	Semiflex padding foam	Behind facing
Automotive consoles	Semiflex padding foam	Behind facing
Automotive dashboards	Rigid structural foam	Glass mat reinforced
Automotive dashboards	Semiflex padding foam	Behind facing

▶

Table 2. Continued

Applications	Type of PU	Remarks
Automotive disc covers	Semiflex struct. foam	RIM
Automotive door leaves	Semiflex struct. foam	(R)RIM
Automotive door linings	Semiflex padding foam	Behind facing
Automotive energy absorption	Flexible foam	Semiflex padding foam
Automotive engine housing	Rigid structural foam	Glass mat reinforced
Automotive external protection	Semiflex structural foam	
Automotive external claddings	Semiflex structural foam	(R)RIM
Automotive external trims	Semiflex struct. foam	(R)RIM
Automotive filters	Rubber	
Automotive filters	Flexible foam	Slabstock, moulded
Automotive handles	Semiflex struct. foam	
Automotive head-rests	Semiflex struct. foam	
Automotive head-rests	Flexible foam	Moulded
Automotive interior trims	Rigid structural foam	
Automotive interior trims	Rigid foam	
Automotive interior paints	Lacquer	Aqueous 2-pack
Automotive interior trims	Adhesive	Aqueous dispersion
Automotive internal protective padding	Semiflex structural foam	
Automotive internal protective padding	Flexible foam	Slabstock, moulded
Automotive knee cushions	Semiflex padding foam	Behind facing
Automotive lever knobs	Semiflex struct. foam	
Automotive lever knobs	Thermoplastic	
Automotive lorry superstructures	Rigid foam	Modular construction
Automotive lorry superstructures	Rigid structural foam	
Automotive mudguards	Semiflex struct. foam	(R)RIM

▶

Table 2. Continued

Applications	Type of PU	Remarks
Automotive off-road cars	Lacquer	External hard top
Automotive repair paints	Lacquer	Water-thinnable
Automotive roof liners	Rigid structural foam	Glass mat reinforced
Automotive roof liners	Rigid foam	Thermoformable
Automotive sealing bellows	Thermoplastic	
Automotive seals	Elastomers	Hot cast
Automotive seals	Rubber	
Automotive seat shells	Rigid structural foam	Glass mat reinforced
Automotive seats	Flexible foam	Moulded, stabstock, flocking
Automotive side impact protection	Flexible foam	Semiflex, energy-absorbing
Automotive sound absorption	Flexible foam	Moulded, flocking
Automotive spare wheel recesses	Rigid structural foam	Glass mat reinforced
Automotive spoilers	Rigid structural foam	Also glass mat reinforced
Automotive spoilers	Semiflex struct. foam	Also in (R)RIM
Automotive spring aids	Elastomers	Hot cast
Automotive spring aids	Rubber	
Automotive steering wheels	Semiflex struct. foam	Cover
Automotive thresholds	Semiflex struct. foam	(R)RIM
Automotive thresholds	Thermoplastic	Also glass fibre reinforced
Automotive tractor cab	Rigid structural foam	
Automotive tractor seats	Semiflex struct. foam	
Automotive underseals	Coating	2-pack
Automotive visors	Semiflex padding foam	Behind facing
Automotive visors	Flexible foam	Slabstock, moulded, padding foam
Awnings	Coating	
Back-drops	Rigid structural foam	"direkt": Nov. 1996
Balls	Flexible foam	Slabstock, moulded

▶

Table 2. Continued

Applications	Type of PU	Remarks
Bath tubs (acrylic)	Elastomers, reinforced	Cold spraying
Bathing fabrics	Fibres	
Bathrooms	Rigid foam	Plumbing blocks
Beer barrels	Semiflex struct. foam	Cover
Bellows	Rubber	
Bicycle saddles	Semiflex struct. foam	RIM
Bicycle saddles	Flexible foam	Moulded
Boats and fittings	Elastomers	Cold spraying
Boats and fittings	Rigid structural foam	
Boats and fittings	Rigid foam	
Bonding foam	Rigid foam	Also "can foam"
Bricks, hollow	Rigid foam	Integrated heat insulation
Building paints	Lacquer	Concrete plaster
Cable sheathing	Thermoplastic	For all types of cable
Carpet underlays	Elastomers	Cold cast
Carpet underlays	Flexible foam	Slabstock, flocking
Carrier bags	Coating	Printable sheets
Cash dispensers	Rigid structural foam	
Chipboard form. free	Binder	Of wood, straw, etc.
Cinema seats	Rigid structural foam	
Clothing	Adhesive	Flexible foam composite
Cold storage cells	Rigid foam	
Cold stores	Rigid foam	
Computer casings	Rigid structural foam	PU steel sandwich panels
Condoms	Thermoplastic	UT June/July 96, p. 3
Construction of industrial buildings	Rigid foam	
Containers of any type	Rigid foam	Oil, water, cereals
Containers, including residential	Rigid foam	Also on-site spraying

▶

Table 2. Continued

Applications	Type of PU	Remarks
Conveyor belts	Elastomers	Cold/hot cast or knife application.
Corrosion prevention	Elastomers	Cold cast, spraying
Corrosion prevention	Lacquer	Surfaces of any type
Corsetry	Fibres	
Couplings	Rubber	
Couplings	Elastomers	Hot cast
Coverings	Rigid structural foam	For industrial plant
Crane buffers	Elastomers	Hot cast
Crude oil tanks	Rigid foam	
Crude oil, see pipelines	Rigid foam	
Cylinders for ind. use	Elastomers	Hot and cold cast
Deburring drums	Elastomers	Hot cast
Dental equipment	Rigid structural foam	
Diaphragms, industrial	Elastomers	Hot cast
Diaphragms, industrial	Rubber	
Diaphragms, industrial	Thermoplastic	
District heating pipes	Rigid foam	
Doors for drinks dispensers	Rigid structural foam	
Doors, gates, heat insulated	Rigid foam	
Drive belts	Elastomers	Hot cast
Duplicating paper	Coating	Microcapsules
Earmarks	Thermoplastic	For (grazing) animals
Electrical insulating paints	Lacquer	
Energy absorbing parts	Flexible foam	Moulded, flocking
Energy-absorbing parts	Elastomers	Hot/cold cast
Energy-absorbing parts	Rubber	
Energy-absorbing parts	Semiflex padding foam	Behind facing
Energy-absorbing parts	Flexible foam	Slabstock, bonded flake
External wall insulation	Rigid foam	Heat skin, panels
Facings	Rigid structural foam	

▶

Table 2. Continued

Applications	Type of PU	Remarks
Facings	Rigid foam	
Facings	Semiflex struct. foam	
Facings	Flexible foam	Slabstock
Fenders	Elastomers	Hot cast, boats, quaysides
Fenders for sports boats	Semiflex struct. foam	
Filter mat carriers	Rigid structural foam	For sewage treatment
Filters for industry	Flexible foam	Crosslinked slabstock
(Fire) hose coatings	Elastomers	Cold knife application, hot curing
Flat roofing	Rigid foam	Panels, moulded foam
Flocking	Adhesive	Electrostatic
Floors for sporting activities, including foamed floors	Elastomers	Cold cast, halls, outdoors
Flower arranging foam	Rigid foam	Hydrophilic, open-cell
Flower arranging substrate	Rigid foam	
Football boot soles	Thermoplastic	TPU studs
Foundry cores	Binder	Foundry resins
Freezers, see "refrigerated . . ."	Rigid foam	
Furniture	Lacquer	
Glass fibre sizes	Binder	Aqueous medium
Grinding discs	Elastomers	Hot and cold cast
Grinding discs	Rubber	
Grips	Semiflex struct. foam	Ski sticks
Gully floors	Elastomers	Cold spraying
Gullys	Rigid structural foam	
Gym mats	Flexible foam	Bonded foam
Heat insulation for floors	Rigid foam	
Hosepipes (also for the fire brigade)	Thermoplastic	
Hospital equipment	Rigid structural foam	

▶

Table 2. Continued

Applications	Type of PU	Remarks
Hospital equipment	Semiflex struct. foam	
Hospital equipment	Flexible foam	Slabstock, moulded
Hot-water cylinders	Rigid foam	
Hydrocyclones	Elastomers	Hot cast
Ice hockey boots	Semiflex struct. foam	
Ice hockey boots	Thermoplastic	
Ice hockey gloves	Semiflex struct. foam	
Industrial filters	Flexible foam	Cross linked slabstock
Inner shoes for skis	Semiflex struct. foam	
Invalid's hoists	Rigid structural foam	
Leather (tanning, dressing)	Coating	Dyeing, impregnation, finishing
Leather for bags	Coating	
Lingerie	Fibres	
Liquefied gas tankers	Adhesive	Insulating panel on LNG tanks
Liquefied gas transport	Rigid foam	LNG cold insulation
Mattresses/cores	Flexible foam	Stabstock, moulded, bonded
Metallic effect	Lacquer	Automotive industry
Moulds	Rubber	
Multistorey car parks	Rigid foam	Panels, moulded foam
Office furniture	Elastomers	Cold cast
Office furniture	Rigid structural foam	
Office furniture	Semiflex struct. foam	
Office furniture	Flexible foam	Slabstock, moulded, flocking
Outer garments	Coating	Dry cleanable
Packaging	Flexible foam	Slabstock, moulded, flocking
Packaging (paper, card)	Coating	Aqueous dispersions
Padding	Rigid foam	Thermoformable

►

Table 2. Continued

Applications	Type of PU	Remarks
Padding	Semiflex struct. foam	For sports equipment
Padding	Flexible foam	Slabstock, moulded, bonded
Paints	Lacquer	Surfaces of any type
Parquet flooring	Lacquer	Sealing, ready-made parquet flooring
Patent leather	Coating	"Cold lacquer"
Petroleum tanks	Rigid foam	
Petrol-pump covers	Rigid structural foam	
Photo-developer accessories	Rigid structural foam	Tanks, rollers
Pipe coatings (ext./int.)	Elastomers	Cold/hot cast
Pipe scrapers	Elastomers	Hot cast
Pipelines	Rigid foam	Petroleum, etc.
Pipes, cold carrying	Rigid foam	
Pipes, heat carrying	Rigid foam	
Pipes, heat carrying	Flexible foam	Slabstock
Pitched roofs	Rigid foam	Panels
Plant containers	Rigid structural foam	
Plastics processing	Adhesive	
Playgrounds (children, school)	Elastomers	Cold cast
Plumbing	Rigid structural foam	
Polishing discs	Elastomers	Hot cast
Polishing discs	Rubber	
Protective clothing	Coating	
Pump linings	Elastomer	Hot cast
Pump stators	Elastomer	Hot cast
Pump stators	Rubber	
Railways	Lacquer	External paints
Rainwear	Coating	
Refrigerated appliances	Rigid foam	Refrigerators, freezers

▶

Table 2. Continued

Applications	Type of PU	Remarks
Refrigerated containers	Rigid foam	Also ships
Refrigerated trucks	Rigid foam	Heat insulation
Refurbishment of old buildings	Rigid foam	
Reinforcements	Rigid structural foam	
Rock consolidation	Binder	Mining, civil engineering
Roller blades	Elastomers	Hot cast
Roller-skate wheels	Elastomers	Hot cast
Roller-skate wheels	Rubber	
Rollers for ind. use	Elastomers	Hot/cold cast
Rollers for ind. use	Rubber	
Rollers for ind. use	Thermoplastic	
Rolling doors, heat insulated	Rigid foam	
Safety glass	Elastomers	Cold cast
Sealant for buildings	Elastomers	Cold spraying
Seals	Elastomers	Hot/cold cast
Seals	Rubber	Also for hydraulics
Seals	Flexible foam	Slabstock, moulded
Seat frames	Rigid structural foam	
Seat frames	Rigid foam	
Seating	Rigid structural foam	
Seating	Rigid foam	Thermoformable
Seating	Flexible foam	Slabstock, moulded
Sheets	Thermoplastic	
Shoe inserts	Elastomers	Hot cast
Shoe inserts	Semiflex struct. foam	
Shoe inserts	Flexible foam	Slabstock
Shoe soles	Adhesive	
Shoe soles	Semiflex struct. foam	Multi-purpose
Shoe uppers	Coating	For leather, textiles

▶

Table 2. Continued

Applications	Type of PU	Remarks
Shoe trees	Rigid structural foam	
Shop fittings	Rigid structural foam	
Shoulder cushions	Flexible foam	Slabstock
Shower cubicles	Rigid structural foam	
Shower trays (acrylic): reinforced	Elastomers	Cold spraying
Shuttering mats	Flexible foam	Slabstock
Shuttering mats	Elastomers	Cold cast
Shutters/shutter boxes	Rigid foam	Hot/cold insulated
Sieves	Elastomers	Hot cast
Sieves	Thermoplastic	Mining, gravel
Skateboard wheels	Elastomers	Hot cast
Ski boots	Thermoplastic	
Skis	Rigid structural foam	Also water skis
Skittle balls	Rigid structural foam	
Sledges	Rigid structural foam	Including bobsleighs
Sound absorption	Flexible foam	Moulded, flocking, low-exp. foam
Sound insulation	Elastomers	Cold cast
Sound insulation	Flexible foam	Slabstock, moulded
Spiral pneumatic sifters	Elastomers	Hot cast
Split leather	Coating	
Sponges	Flexible foam	Slabstock
Spring elements	Elastomers	Hot cast; solid, foamed
Stable-venting elements	Rigid structural foam	
Stockings, socks	Fibres	
Stoving lacquers	Lacquer	
Surfboards	Elastomers	Cold cast
Surfboards	Rigid structural foam	
Surfboards	Rigid foam	

▶

Table 2. Continued

Applications	Type of PU	Remarks
Swimming pool cleaning equipment	Rigid structural foam	
Synthetic leather	Coating	Textile backing mats
Table surrounds	Elastomers	Cold cast
Television casings	Rigid structural foam	
Tennis training walls	Flexible foam	Slabstock
Tent roofs, floors	Coating	
Textile knitted fabrics	Fibres	
Textile ribbons	Fibres	
Textile woven fabrics	Fibres	
Textile yarns	Fibres	
Textiles, laminated	Flexible foam	slabstock, web form
Thermal containers for camping	Rigid foam	Hot/cold insulated
Toothed belts	Elastomers	Hot cast
Toothed belts	Rubber	
Toothed belts	Thermoplastic	
Toys	Rigid structural foam	
Traffic light casings	Rigid structural foam	Euro-Tunnel
Trouser fabrics	Fibres	
Tubbing rings	Rigid structural foam	
Vacuum cleaner casings	Rigid structural foam	
Wall heat insulation	Rigid foam	
Water heaters	Rigid foam	
Window seats	Rigid foam	Hot/cold insulation
Windows	Rigid structural foam	

4 Why polyurethanes?
List of properties

In the previous chapters, we have asked whether we really need polyurethane and, if so, what for. The answer is an unlimited range of applications which sets it apart from other materials. In the next few chapters, we will look at the reasons for the universal acceptance of polyurethanes. The most obvious reason for using a material lies in its macroscopic properties. In the "Polyurethane Age" (Fig. 13) we cannot speak of definitive properties of polyurethane; there are at least as many properties as forms. However, they can be summarized as follows (Table 3):

Table 3. General properties of PU

	Hard	Soft	Flexible	Rigid	Elastic	Light
PU foams	+	+	+	+	+	+++
Non-cellular PU	+	+	+	+	+	+

Other, or rather *not the real* origins of the Polyurethane Age lie not in process engineering but in the molecular structure, i.e. in the chemistry of polyurethanes (Chapters 5 and 6).

4.1 Physical and chemical properties according to forms

PU foams are lightweight materials, 98 to 99 % of which consists of air or – in the case of closed-cell foams – of gaseous blowing agents and less than 5 % of solid cell walls. Their mechanical properties essentially depend on their density. Table 4 shows the densities of PU foams between air, water and other materials.

Flexible foams are open-celled and have relatively low resistance to deformation under pressure (DIN 7726).

Flexible slabstock foams are classified as

      polyether foams and
      polyester foams,

Table 4. Densities of air, water and other materials

Description	Density in kg/m^3
Dry air	1.3
Lightest PU (flexible) foam	about 10
Cork	200 to 350
Dry wood	400 to 800
Heaviest PU (structural) foam	800 to 900
Water	1000
Magnesium	1740
Aluminium	2702
Titanium	4510
Structural steel	7850

according to their different chemical composition. Polyether foams predominate in terms of quantity at about 94 %. They, in turn, can be distinguished as follows:

Standard foams are prevalent in terms of quantity but special types are becoming increasingly important:

Highly elastic foams are also described as "high resilient" (HR) or "cold stabstock" foams.

Flame retardants containing HR types also go by the name of "combustion modified", or *"CMHR" foams.*

High load bearing foams are known as *HL foams.*

Supersoft and *special foams*, for example *flame-laminated and/or high frequency (HF) welded foams*, complete the range.

Flexible foams generally have densities of 20 to 40 kg/m^3. Extremely heavy qualities up to 130 kg/m^3 are available for special fields of application.

As the density increases, the suitability for demanding dynamic applications is improved. Compressive strength and indentation resistance are important characteristics. They are determined from the compressive deformation graph (Fig. 34) which provides information about the hysteresis area and at the same time about the elasticity and (mechanical) energy absorption of the flexible foam. The values for compression set, tensile strength and elongation at break are other criteria (Table 5).

Slabstock foams are also stable in dry heat (83 °C/3 months). Their thermal conductivity is between 0.030 and 0.040 W/m K at 0 to 70 °C. They are permeable to X-rays but impermeable to UV and IR radiation. They are an ideal material for airborne sound insulation owing to their open cell structure and elastic properties.

The properties of *flexible moulded foams* depend, in particular, on the chemistry and processing of the respective PU systems. A distinction is made between *hot-cure foam* and *cold-cure foam*, depending on whether

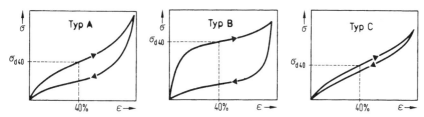

Fig. 34. Compressive stress-deformation graphs of flexible PU foams
Mechanical hysteresis: ▶= stress application curve,
◀= stress removal curve
σ_{d40} compressive stress at 40 % compression (compressive strength)
Type A: flexible PU foam with normal energy absorption
Type B: flexible PU foam with high energy absorption
Type C: flexible PU foam with low energy absorption

Table 5. Mechanical properties of flexible slabstock foams

Flexible slabstock foam	DIN 53420 Density kg/m³	DIN 53577 Compressive strength kPa	DIN 53572 Compression set %	DIN 53571 Tensile strength kPa	DIN 53571 Elongation at break %
Polyester foam	35	6.5	4 to 15	160 to 220	100 to 450
Polyether foam (standard)	36	4.5	2 to 10	100 to 180	100 to 400
Polyether foam (HL)	36	6.0	*	*	*
High resilient foam (HR)	34	2.7	2 to 10	50 to 120	80 to 200

* = highly dependent on formulation

Table 6. Mechanical properties of flexible moulded foams

Flexible moulded foam	DIN 53420 Density kg/m^3	DIN 53577 Compressive strength kPa	DIN 53572 Compression set %	DIN 53571 Tensile strength kPa	DIN 53571 Elongation at break %
Hot-cure foam I	33	4.0	4	90	190
Hot-cure foam II	33	4.7	5	112	190
Cold-cure foam I	38	3.3	4	110	125
Cold-cure foam II	37	4.0	6	150	150
Cold-cure foam III	44	5.6	5	152	150

Hot-cure foam I: Polyether OHZ = 56/TDI 80   Cold-cure foam I: modified TDI
Hot-cure foam II: Polyether OHZ = 35/TDI 65   Cold-cure foam II: TDI/MDI mix
                                                                             Cold-cure foam III: MDI

the foam part has to be baked in the mould or can be demoulded while cold. The reason why it is still hot will be explained more fully in Chapter 6. Table 6 illustrates the influence which the raw material base has on flexible slabstock foams.

If a property is improved particularly markedly in a flexible moulded foam, this is frequently at the expense of other properties; this fact has to be allowed for when designing seat cushions for individual applications (furniture, cars). Fig. 35 shows the excellent mechanical damping

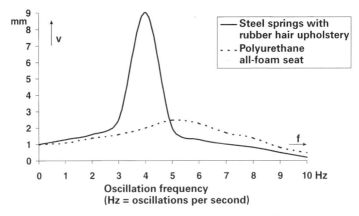

Fig. 35. Oscillation behaviour of various seat designs: the PU all-foam seat reduces strain on the driver and therefore contributes to safety

permitted by flexible moulded foam in the example of an all-PU foam car seat.

So-called *cellular PU elastomers* are high-performance, elastically deformable flexible moulded foams with densities ranging from 300 to 600 kg/m^3.

However, the chemistry and process engineering as well as the properties and fields of application of cellular PU elastomers differ so fundamentally from those of the "genuine" flexible moulded foams discussed above that they are generally considered as PU elastomers in practice (see below).

Semi-rigid padding foams are completely different from flexible foams owing to their much higher functional density, compression strength and compression set (Table 7).

Table 7. Mechanical properties of semi-rigid padding foams

Semi-rigid padding foams	DIN 53420 Density kg/m^3	DIN 53577 Compressive strength kPa	DIN 53572 Compression set %	DIN 53571 Tensile strength kPa	DIN 53571 Elongation at break %
More flexible type	130	45	<15	150	40
More rigid type	150	95	30	350	50
Energy-absorbing foams	80 to 100	150 to 320	<45	350 to 450	10 to 25

Their excellent mechanical damping behaviour (Fig. 36) makes them eminently suitable for protective cushions in car interiors. A variation of semi-rigid padding foams, known as EA (energy-absorbing) foams, protects passengers both in the interior as side impact protection and externally as bumper material.

Rigid foams have relatively high resistance to deformation under pressure (DIN 7726). The compressive strength (DIN 53421) depends on the direction of foaming owing to the anisotropy of the "actual" rigid foams.

A dodecahedron with 12 equilateral pentagons can be considered as a "cell window" for the "ideal" rigid foam cell which can be in compressed and extended form in the "actual" foam (Fig. 37).

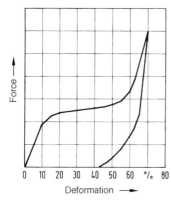

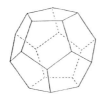

Fig. 36. The size of the hysteresis loop in the force-deformation graph demonstrates the good damping behaviour of semi-flexible PU padding and moulded foams

Fig. 37. Ideal rigid foam would be isotropic and, with uniformly dodecahedral cells (left), would have identical properties in all directions; actual foam has anisotropic cells with compressed (centre) or elongated (right) cell windows and directional properties

The main properties of rigid foams are their particularly low heat conduction owing to the insulating gases trapped in the closed cells and the ability to form composites with almost all flexible or rigid facings. Densities invariably depend on the given application and range from 30 to $90\,kg/m^3$, for example in the case of slabstock foam, and between 30 and $40\,kg/m^3$ in the core of laminated foams. The density of in-situ foam is 37 to $40\,kg/m^3$, of sprayed foam about $55\,kg/m^3$ and in PU-insulated district heating pipes not below $80\,kg/m^3$. A minimum density of $30\,kg/m^3$ is recommended for low-temperature insulation in refrigerators and freezers. Table 8 shows the properties as a function of density in the example of continuously produced PU rigid foam blocks. Obviously these values cannot easily be transferred to other applications.

PU rigid foam can be used at between $-200\,°C$ and $+150\,°C$; it can withstand short-term stress of up to $250\,°C$, of the type occurring, for

Table 8. Properties* of continuously produced rigid slabstock foams

Density according to DIN 53420	kg/m^3	32	50	90
Compressive strength II**	MPa	0.20	0.35	0.70
according to DIN 53421 I***	MPa	0.11	0.20	0.60
Tensile test according to DIN 53430:				
Tensile strength	MPa	0.20	0.27	0.9
Elongation at break	%	4.0	5.2	5.0
Tensile modulus of elasticity	MPa	6.2	6.5	29.2
Dimensional stability**** according to DIN 53431:				
3 h at −30 °C		−0.3	−0.1	−0.1
5 h at +130 °C		+2	+0.5	+0.2
Open-cell character according to ASTM-D 1940-42 T	%	9	12	8
Thermal conductivity according to DIN 52616	W/m · K	0.021	0.021	0.027

*   Properties obtained under optimum processing conditions
**   Compressive strength, measured in expansion direction
***   Compressive strength, measured perpendicularly to expansion direction
****   Given as percentage change in volume

example, when PU sheets are laid in hot bitumen. It is odourless and rot-proof, resistant to the plasticizers which may be present in covering film and to fuels, mineral oils, dilute acids and alkalis. About 40 years ago, it was stated in a "predecessor" of this book, that PU foams are "thermoformable", despite being crosslinked (Fig. 38).

This property can be improved by developing suitable formulations so sheets cut from appropriate slabstock foam or produced continuously can be thermoformed at 190 °C, together with a decorative layer, to form car roof liners. "Cold formable" systems can also be processed at 90 to 130 °C in a variation of the process.

Whereas flexible and rigid foams differ fundamentally in chemical structure (slightly/highly crosslinked), *structural foams* are subjected to different process engineering: they can be obtained as both flexible and rigid products merely by foaming in the mould (see Section 5.2). According to DIN 7726, "structural foam" is the name for (PU) mouldings which have a cellular core and a non-cellular outer layer of the same material and are produced in a single pass. The transition

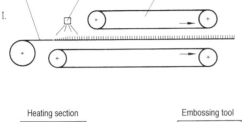

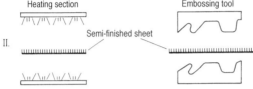

Fig. 38. Special PU rigid
foams are thermoformable

from the non-cellular edge region into the cellular core is, as in nature
continuous (Fig. 39) rather than abrupt (Fig. 40).

The properties of structural foams are derived from this sandwich
structure.

Chemistry and process engineering also affect the final properties of the
ready-to-use moulding, enhanced by the glass fibre reinforcement, so we
can only give a general overview at this point. Furthermore, some

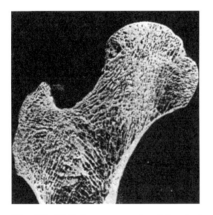 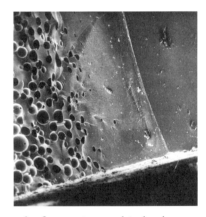

Fig. 39. Cellular core and solid exterior; examples from nature and technology:
cross section through bone and electron microscope photograph of the edge region
of a PU structural foam

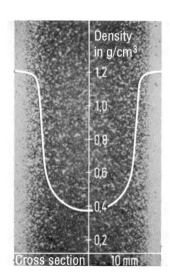

Fig. 40. Structural foam: density trend
over cross section of sample

parameters are more dependent on application: for example, abrasion-resistance is more important than low-temperature notched impact strength in the case of a shoe sole made of flexible PU structural foam. The reverse applies to a vehicle body part made of microcellular or solid polyurethane.

Table 9 summarizes some empiric values for the properties of flexible structural foam, microporous and solid polyurethanes.

The influence of glass fibre reinforcement on the properties can be demonstrated by a simple comparison: whereas the flexural modulus of elasticity is a few 10 to 10^2 MPa at room temperature (according to ASTM-D 790) in the case of ductile PU structural foams of average density ("overriders" in Table 9), flexible microcellular and solid polyurethanes, 22 % of short glass fibres in PU systems raise this modulus to 1300 MPa, for example in the case of car body parts.

Mouldings made of rigid structural foams are light and robust while being convenient and attractive. However, Table 1 only provides a small selection of products with useful properties. Table 10 shows features which might be helpful for creative designers.

Rigid structural foams can be used at up to 120 °C and as PIR systems (polyisocyanurate, see Chapter 6) even at higher service temperatures.

Table 9. Mechanical properties of flexible structural, microporous and solid polyurethanes

Material	DIN 53420 Density kg/m³	DIN 53506 / DIN 53505 Shore A / Shore D	DIN 53571 Tensile strength kPa	DIN 53571 Elongation at break %	DIN 53507 Tear propagation resistance kN/m	DIN 53561 Abrasion resistance mg	DIN 53577 Compressive strength %	DIN 53572 Compression set %	DIN 53453 Notched impact test kJ/m²
Structural foam Shoe sole/ polyester	50	60	10 000	500	8	<60	–	–	–
Structural foam Bicycle saddle Skin:	145	–	400 / 2600	170 / 155	– / 5		32	<15	– / –
Structural foam Overriders	70	75	5700	230	12	–	–	–	no breakage
Microporous PU Lorry mudguard	1000	54	22	180	51	–	–	–	no breakage
Solid PU External vehicle body parts	1050	45	22	260	52	–	–	–	no breakage

Table 10. Mechanical properties of rigid structural foams and microporous polyurethanes

Properties	Unit	DIN test	Materials		
Density	Kg/m³		400	600	1050
Flexural strength	MPA	53432	25	45	–
Tensile strength	Mpa	53432/53455	8	18	50
Elongation at break	%	53432/53455	7	7	15
Flexural modulus	Mpa	53432/53455	600	1050	2000
Tensile modulus	Mpa	53432	305	600	–
Impact strength	kJ/m²	53432/53455	–	–	55

In *solid PU elastomers*, the properties, like those of structural foams, are highly dependent on chemical composition and process engineering, in particular on the method of shaping. Over the entire range, solid PU elastomers have two basic properties in common: high elasticity regardless of hardness and high resistance to wear under different stresses. With reference to the modulus of elasticity, Fig. 41 compares PU elastomers, including glass fibre reinforced grades with other

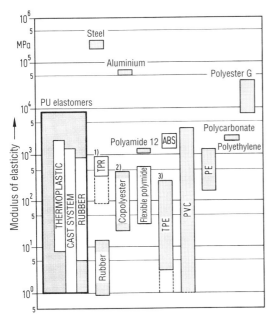

Fig. 41. Moduli of elasticity of PU elastomers compared to other materials
1) TPR = thermoplastic rubber
(EPDM + polypropylene)
2) Copolyester = polybutylene terephthalate polytetraoxymethylene
3) TPE = thermoplastic elastomers, styrene butadiene block copolymers and (sometimes) polypropylene

materials. *Solid PU cast elastomers* are available with a Shore hardness ranging from 25 A to 70 D and have a modulus of elasticity of 4 to 700 MPa. Table 11 shows properties of a PU hot-cast elastomer (Vulkollan) which has been around for over half a century.

Table 11. Properties of PU elastomers (hot-cast system based on NDI) with various polyols

Properties	Unit	Test standard	Elastomer type		
			a	b	c
Shore A hardness		DIN 53505	88	89	90
Ultimate tensile strength	Mpa	DIN 53504	45	47	24
Elongation at break	%	DIN 53504	695	679	503
Tear propagation resistance	kN/m	DIN 53515	72	42	23
Impact resilience					
Abrasion	%	DIN 53512	54	60	68
Compression set (at 70 °C)	mg	DIN 53516	40	32	35
	%	DIN 53517	26	20	20

Whereas the rigidity is almost constant, the mechanical properties clearly depend on the type, i.e. on the chemical composition. PU cast elastomers do not contain plasticizers and only swell slightly in mineral oils. Some types have to be protected from hydrolysis by suitable additives (Section 6.3.4), but they are resistant to light, oxygen/ozone and UV radiation.

In contrast to conventional UP and EP systems, *PU electrical embedding compounds* exist in flexible to rigid variations with a glass transition temperature range of $-40\,°C$ to $130\,°C$. So-called *EP/IC moulded materials* with a $T_G = 300\,°C$ and long-term thermal stability according to IEC 216 of $198\,°C$ are a special case. In contrast to UP, EP and PU resins, moreover, they attain the classification V-O at 3.2 mm in the UL 94 fire test.

PU sealants are competing with known materials such as polysulphides and silicones; they are available as firm and self-levelling sealants with Shore A hardnesses of 10 to 45 or 20 to 60.

Thermoplastic PU elastomers (TPU) combine the properties of PU elastomers with the economic processability of thermoplastic polymers.

The most important mechanical properties are summarized in Table 12 which relates to a commercial range of products. These properties depend on the raw material basis, as is typical of PU (polyesters, ethers, etc., see Section 6.3.2). TPUs are free of plasticizers over all hardness ranges.

Table 12. Mechanical properties of thermoplastic polyurethanes (TPUs) [73]

Properties	DIN test	Unit	Range of values
Shore A hardness	53505	–	77 to 98
Shore D hardness	53505	–	28 to 73
Modulus of elasticity	53455	MPa	10 to 650
Ultimate tensile strength	53504	MPa	26 to 50
Elongation at break	53504	%	250 to 740
Abrasion loss	53516	mm^3	20 to 60
Rebound resilience	53512	%	30 to 48
Tear propagation resistance	53515	kN/m	40 to 250
Compression set 70 h/22 °C	53517	%	17 to 30
Compression set 24 h/70 °C	53517	%	32 to 65
Density	53479	g/cm^3	1.11 to 1.25

PU rubbers have excellent resistance to oil and fuel and particularly high gas impermeability in addition to the high abrasion resistance and tensile strength typical of PU elastomers. They maintain their flexibility down to −40 °C and are thermally stable to +125 °C. Peroxide- and sulphur-crosslinked types are available in hardness ranges of 45 to 80 Shore A and isocyanate-crosslinked PU rubbers between 70 Shore A and 50 Shore D. Table 13 compares some striking properties of PU rubber with those of other special rubbers.

PU paints, also in the form of water-thinnable paints and powder coatings, combine two groups of properties: mechanical properties of the lacquer film and its chemical resistance to aggressive environmental influences (light, weather, etc.). The chemical composition allows adjustments from tough as horn to rubbery elastic and permits processing down to the 0 °C limit. The paints are known for their high scratch, abrasion and impact resistance and they generally adhere very well to a wide variety of substrates. The excellent gloss retention of two-pack polyurethane paints/HDI base (see Section 6.3.1.) under

Table 13. Comparison between properties of PU rubber and other special types
(Source: Rhein-Chemie, Mannheim)

CSM: Chlorosulphonated PE rubber   ACM: Acrylate rubber
VMQ: Vinyl methyl silicone rubber   FPM: Propylene tetrafluoroethylene rubber
NBR: Nitrile butadiene rubber   ECO: Epichlorohydrin copolymer rubber

	Mechanical strength	Abrasion resistance	Ozone resistance	Oil resistance	Service temperature °C
PU	1	1	1	1	125
CSM	3	3	1	2	120
ACM	3	3	1	3	150
VMQ	4	4	1	4	180
FPM	2	3	1	1	230
NBR	2	2	3	2	110
ECO	3	3	2	5	130

1 = very good 2 = good 3 = average 4 = satisfactory

weathering influences is compared with other lacquer binders in Fig.
42. A three-layered structure comprising zinc primer, intermediate and
top coat provides exceptional protection against corrosion and is
particularly resistant to corrosive condensation. Moisture-curing one-
pack systems are preferred for taking maximum abuse. Aqueous PU
systems can help to reduce solvent emissions during painting. At
present they have a market share of only 8 %, but this will increase
to 25 % in the next 5 years. Fig. 43 shows the emissions generated
during conventional painting in comparison with those using aqueous
PU paint systems.

PU coatings and finishes for textiles, paper and leather show how
the lacquers adhere well to substrates and are highly flexible even at
low temperatures. In contrast to cheaper coatings composed of SBR,
acrylate, nitrocellulose or PVC, PU coatings frequently have higher
abrasion resistance, tensile strength and tear propagation resistance and
are free from plasticizers. In addition, embrittlement due to light and
long-term storage can be avoided. Polyurethane is also the most
important polymer for *poromeric synthetic leather* (for example,
Corfam). It looks like leather and has steam permeability similar to
that of the natural product. However, poromers obviously lose strength
as their porosity increases and do not yet have the properties essential
for comfort in wear which natural leather has.

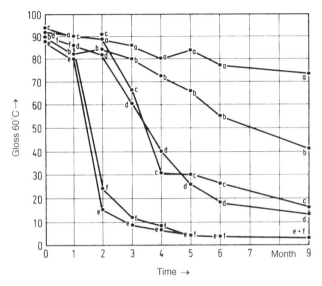

Fig. 42. Relative gloss retention of various paint binders in outdoor conditions
(Florida test)
a: 2-pack PU paint, HDI basis; b: alkyd resin; c: chlorinated rubber; d: 2-pack PU
paint, TDI basis; e: urethane oil; f: 2-pack epoxide resin lacquer

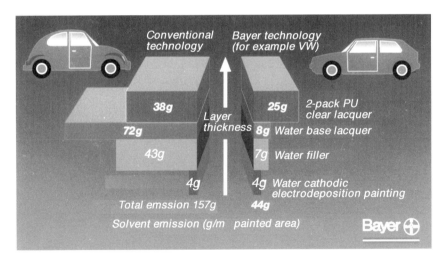

Fig. 43. Solvent emission is reduced by 72 % by using water/PU paint systems
during the cathodic electrodeposition painting of cars rather than conventional
methods

The adhesive strength of polyurethanes *"in statu nascendi"* on the surfaces of all materials is utilized in PU adhesives. Chemistry again enables the adhesiveness to be controlled. Hence, the resistance of the bonded joint to heat, water, solvents, plasticizers, grease, oils and ageing generally increases with the crosslinking density (see Chapter 6). The bonded joint is frequently firmer than the adherend (Table 14).

Table 14. Shear tensile strength of multi-phase adhesives (overlap length 10 cm, curing 15 min/120 °C)

Material	Shear tensile strength at 23 °C	
	Adhesive 1 MPa	Adhesive 2 MPa
Polyester, unsaturated (SMC)	9.7 F	9.1 F
Polybutylene terephthalate	9.8 F	–
Polycarbonate	12.0 K	–
Plastic blend PC/ABS	11.2 K	–
Sheet steel (pretreated by cathodic electrodeposition painting)	14.4 F	–
Glass (Duran)	4.4 F	–

Boards made of renewable raw materials (chipboard) and PU rigid foam scrap provide examples of the properties of materials produced from scrap and *PU/PMDI binders* (see Section 6.3.1). Table 15 shows that all the above-mentioned agricultural raw materials, apart from three exceptions, not only attain but surpass the V20 quality for wood according to DIN 68763 and the V100 classification. Boards produced from shredded rigid foam scrap bonded by polyurea have the following property profile (Table 16).

The elastic properties of *PU fibres (elastanes)* and their high extensibility have almost completely displaced rubber threads (elastodienes) as elastic fibres in the textile industry. The reason for this is that "the chemistry is right" (segmented molecular structure, see Section 6.4). Process engineering also helped PU fibres: whereas rubber threads are available only as monofilaments, elastanes are produced exclusively as continuous yarns (filaments) with finenesses of 1 to 500 tex (g/1 km). They have elongations of about 420 to 800 % and strengths – based on original fineness – of about 4.5 to 12 cN/tex, depending on type.

Table 15. Properties of polyurea-bonded chipboard produced from various renewable raw materials (density about 630 kg/m^3)

Raw material	PMDI % a.a.	Shear tensile strength* MPa		Flexural strength** MPa
		V20	V100	
Standard values according to DIN (wood)	–	0.35	0.15	18
Pine	6	0.95	0.27	28
Barley straw	9	0.74	0.21	31
Oats straw	9	0.60	0.22	34
Lupins	9	1.10	0.24	24
Bagasse	6	0.90	0.34	28
Flax shives	7	1.13	0.16	21
Coir	8	1.02	0.45	–
Alfa grass	8	0.60	–	15
Sisal	9	0.35	0.05	15
Bamboo	9	0.80	0.24	12
Rice husks	9	0.38	0.05	13
Peanut shells	9	0.62	0.02	7

*   DIN 53265 test
     – V20: without pretreatment; dry
     – V100: 2 h boiling; wet (DIN 68763)
** DIN 52362 test

Table 16. Properties of boards made from PMDI-bonded rigid foam scrap

Density	about 600 kg/m^3
Swelling in thickness (24 h in water/20 °C)	1 to 4 %
Flexural strength (with paper coating)	15 to 20 MPa
Flexural strength (without paper facing)	20 to 24 MPa
Thermal conductivity	0.06 to 0.08 W/(m K)

PU microcapsules can have diameters of 1 to 5000 m and wall thickness of only 0.03 to 0.05 m. Capsule diameters of 5 to 8 m are used for duplicating paper.

PU gels are molecular networks capable of swelling in water. They absorb up to 90 % of water which cannot be squeezed out.

Organo mineral products, also in foamed form with a high density, have low heat of combustion and high fire endurance; open-cell lightweight foams have densities between 12 and 20 kg/m^3. The property profile as a whole has not been adequate for significant use in the past.

4.2   Fire behaviour

According to a group of experts appointed by the EU Commission, fire performance is *not* a material property. In addition to many other criteria, it also depends on the breakdown, i.e. the surface of the material: heating has to be carried out gently to ignite a solid 1 kg wooden block, whereas a single match and fractions of a second will transform 1 kg of wood shavings into an inferno. We know that solid iron does not burn but if it is finely divided (pyrophoric iron), it glows on contact with atmospheric oxygen. Obviously, many inorganic materials are incombustible and all organic materials are combustible: including polyurethanes. A PU with a large surface area, for example an open-cell flexible foam, burns more rapidly than a solid part of the same weight.

There is therefore a risk of fire in both production and use of PU. This does not arise during storage and handling of the raw materials on account of their relatively high flash points, but the risk of spontaneous combustion must be allowed for, for example during *production* of foam slabstock: with inaccurate metering, the pronounced exothermic character of the PU reaction can lead to scorching in the core of the block and even to spontaneous combustion. The risk is averted by storing freshly foamed slabstock in fireproof rooms until it has completely cooled (see Section 5.2) – and this also helps with regard to insurance. The fire precautions adopted for flammable industrial products also apply to the storage of other PU finished products.

Polyurethanes can be provided with a flame-resistant finish for the respective *applications* (see Section 6.3.4.), which are classified in accordance with fire safety tests. Secondary effects of fire are also assessed, for example burning drips, flue gas density and toxicity. Fire tests are not internationally standardized and usually vary from country to country. Table 17 shows the diversity of fire regulations as applied to PU rigid foam for building purposes, in various countries. They also depend on the individual application of the PU. For example, specific fire tests and regulations are adapted, for example, to mining and transport, furniture and fittings and, in particular, to the building industry.

Table 17. Fire regulations governing flame-resistant rigid PU foams in building materials and flame-retardant systems

Country	PU	. . . with facings such as Al, steel sheet, asbestos cement or plasterboard	. . . with PIR structures	. . . with PIR structures and facings, for example Al foil
Federal Republic of Germany, DIN 4102	B2	B1	B2/B1	B1
Belgium/Epiradiateur	M3/M4	M1	M2/M1	M1
France/Epiradiateur PPF test*	M4	M1	M2/M1	M1
Great Britain BS 476				
Ignitability	X	P	P	P
Fire propagation (I/i_1)	44/34	<10/<10	12–20/7–10	5/about 3
Surface spread	IV	I	II(I)	I
Netherlands NEN 3883				
Uitbreiding	V/IV	I	II/I	I
Overslag	III	I	II/I	I
Switzerland	V/I	V/3	V/1–3	V/3
Scandinavia				
Panel test (Brown box. comparison to wood, Schlyter test,	worse	better	better	better
flash-over)	yes	no	no	no
USA: tunnel test				
ASTM-E 84, rating	≫500	≤25	>25 (≤25)	≤25

* Flame propagation capacity

Particular attention is paid to fumes. While the *smoke density* can be reduced by special flame retardants in polyurethanes, so exits can be seen, it is difficult to influence the *toxicity of the fumes*. Like any nitrogen-containing organic substance (for example, wool), polyurethanes give off steam, carbon dioxide, carbon monoxide, nitrogen oxides, traces of hydrogen cyanide and sometimes even isocyanate vapours in addition to carbon black (see above). The risks associated with fumes generally depend on the quantity of burning material, the size of the room and ventilation, so analytical data on fumes produced in the laboratory are inconclusive with regard to the health hazards encountered in an actual fire: every fire is different! However, standard tests in the laboratory and on animals are essential for comparing various materials. The test data for PU foams in current use shown in Fig. 44 convincingly illustrate that their fumes are no more toxic than those given off by natural substances such as burning wood, cork, wool or cotton. Wood and cork evolve lethal gases at much lower temperatures than polyurethanes, for example in the early stages of a fire. It can be assumed that the toxicity of the fumes given off by PU foams also applies to other polyurethane materials. Although the chemical composition influences the physical properties of polyurethanes, they also have many chemical parameters in common.

Flexible slabstock foams with organic and inorganic flame retardants, sometimes in high concentrations, satisfy stringent fire regulations (California Bulletin 121, 133). They are known as "CMHR" (combustion modified high resilience) foams and, in the form of upholstery and mattress foams, meet the higher British requirements concerning ignitability. Flammability can also be reduced by applying external flame-retardant dispersions according to application. *Flexible moulded foams* (*cold-cure foam*) without additional flame retardants generally meet the requirements of the fire test MVSS-302 applicable to automobile trims, but not *hot-cure foam*. *Rigid foams* with a flame-resistant finish can be classified in the B2 category of building materials (normally flammable) according to DIN 4102. Poly-isocyanurate (PIR)-modified rigid foams (see Section 6.1) with an additional flame-retardant finish fall under classification B1 (flame retardant).

Rigid structural foams with flame retardants fall within the classification VO or 5V according to UL Subject 94, a fire test recognized in the USA electrical industry (Underwriters Laboratory). The VO classification of EP/IC casting resins has already been referred to above. With PMDI, i.e. polyurea-bonded chipboard, the flue gas density in the event of a fire is similar to that of boards bonded in the conventional way with UF or PF; however, the toxicity of inhaled flue gases is lower.

Care has to be taken when interpreting *test* results of polyurethane fire safety tests: the results do not provide definitive evidence about every fire risk which might be encountered; as mentioned above, every fire is different!

	Material	T_v °C	LC_{50} g/m^3	LC_{50} cm^3/m^3
PU	Rigid foam (Density = 40 kg/m^3)	600	≥ 6.6	$\geq 165^*$
	Flexible foam (Density = 30 kg/m^3)	600	≥ 19	$\geq 666^*$
Wood	Spruce (Density = 500 kg/m^3)	600	≥ 28	$\geq 54^*$
	Fir (Density = 500 kg/m^3)	600	≥ 27	$\geq 54^*$
	Pine (Density = 500 kg/m^3)	600	≥ 24	$\geq 48^*$
		500	≥ 18	$\geq 36^*$
Wool		600	≥ 6	–

* Higher LC_{50} values are obtained with lower densities

Fig. 44. Toxicity of PU fumes in comparison with wood and wool (LC_{50}-values according to DIN 53436)

4.3   PU and health

Material exists for man: not the other way round! This applies to polyurethanes which can improve our health when used in the appropriate way. We happily sleep and walk on them, they keep food fresh, we drive and travel safely and comfortably with PU, healthy sport is played with PU equipment, etc. Fully reacted polyurethanes are chemically inert and do not contain biologically available isocyanate groups (see Section 6.3.5). This applies in particular to polyurethanes which come into contact with food and to commodities covered by national legislation. Even toys have to comply with legal regulations. Migration tests carried out on finished PU articles demonstrate that the prescribed limits for Sb, As, Ba, Cd, Cr, Pb, Hg and Se are not exceeded.

Flexible slabstock foams with a defined composition were tested on skin and on animal feeds. No harmful effects were observed, so the foams can be considered to be physiologically safe. Like all inert fine dust, PU fine dust can cause coughing and physical (facial) skin irritation above the MAC value of 6 mg/m^3. PU carpet backings are odourless and, in contrast to latex, do not contain components which can be extracted by water. This has a welcome side effect: there are no water marks! The Federal Environment Office noted, after evaluation of all data on chipboard produced using PMDI, i.e. bonded with polyurea, that there is no reason to ban the use of PMDI. The product and therefore the chipboard are free from formaldehyde and are suitable for the storage and transport of fruit and vegetables, according to the Federal Health Authority.

4.4   An overview of tests

The polyurethane age owes its existence to the wide range of properties. Some of them can be determined quite pragmatically and inexpensively: its "perkiness" could be determined empirically by dipping a thumb into the foam and observing how quickly the foam "recovers". Nowadays, a compressive stress deformation graph is obtained by the standardized compression test using electronically controlled recording and measuring apparatus. Stress application and stress removal curves define the mechanical hysteresis area and determine whether the foam is "perky" or "tired" (cf. Fig. 34). Another example of modern measuring technology involves recording the "mechanical spectrum" of polyurethanes, which used to take several days or weeks. The shear modulus and the mechanical damping of the plastic are measured as a function of the temperature in the torsion pendulum test. After the sample has been fixed and cooled for about half an hour to minus $150\,°C$ at about $1\,K/min$, a computerized, fully automatic system (Fig. 45) measures the vibrations and records the shear modulus and damping curves over a temperature range of $-150\,°C$ to $+150\,°C$ within 5 hours. The "mechanical spectra" show not only the temperature at which the material is elastic or brittle (Fig. 46) but also connections between the morphological structure (see Section 6.4) of the material and its macroscopic properties.

Without providing further details of apparatus, the purpose of *short-term laboratory tests on standard specimens*, is to obtain reproducible characteristics on the one hand for research and development and on the other hand for application engineering and quality control. The modern testing of PU is adapted mainly to its *application*; for example

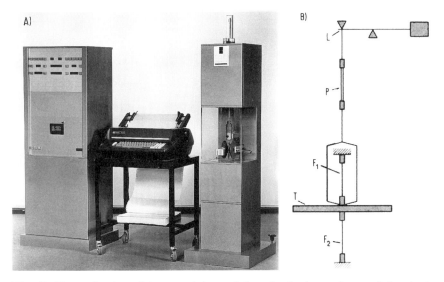

Fig. 45. Determination of shear modulus and damping in the torsion pendulum test
A) General view of a fully automatic system (computerized)
B) Diagram of the experimental set-up
L Compensation for sample length, P specimen, $F_{1/2}$ spring, T mass of inertia

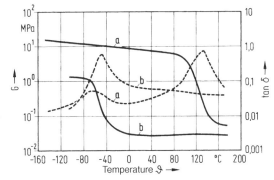

Fig. 46. Example of shear modulus and damping as a function of temperature of
a: rigid PU foam
b: flexible PU foam
——shear modulus G
- - - - tan δ

"rigid foam" is tested not as such but as the PU rigid foam for refrigerators and freezers. Table 18 shows the test criteria for this purpose.

Designers must be familiar with the fatigue behaviour of the material in use. *Sustained stress cycle* and creep tests can be carried out. Special specimens or finished parts are used for this purpose. In the first case,

Table 18. DIN/EN/ISO standards for testing polyurethanes as a function of application: PU rigid foam for refrigerator insulation in this case (Source: Bayer AG)

Density	DIN EN ISO 845
Stretch or tensile test	DIN 53 430
Compressive strength test	DIN 53 421
Bending test	DIN 53 423
Dimensional stability	DIN 53 431
Tensile test (in smooth area)	DIN 53 292
Shear strength	DIN 53 427
Open and closed cells	DIN ISO 4590
Steam permeability	DIN 52 615
Water absorption	DIN 53 433
Heat conduction (heat flow meter)	DIN 52 616
Heat conduction (shielded hotplate)	DIN 52 612
Dimensional stability (under heat exposure and bending stress)	DIN 53 424
Torsion pendulum test	DIN EN ISO 6721
Creep test	DIN 53 444

the specimen is loaded with a defined number of stress cycles. In the fatigue test according to DIN 53574, for example, a PU flexible foam cushion is stressed and released 80 000 times with a mass of 75 kg at room temperature. The hardnesses are then determined by the convention indentation test. The results comply quite well with known values. In contrast to the sustained stress cycle test, the specimen is stored for a prolonged period under a predetermined constant stress (tension, pressure, etc.) in defined climatic conditions in the creep test and changes in properties are observed at time intervals. Fig. 47 shows approximate results and evaluations of creep tests.

Realistic test periods range from at least 100 hours to a few years.

Laboratory tests are frequently followed by complex *tests on the finished part*, whose geometry and possibly inserts and/or facings not made of PU ultimately determine the functional properties of the polyurethane or polyurethane-containing finished part. This will be illustrated by two examples:

Despite quite realistic flexural fatigue tests (see above) on the entire shoe with a foamed-on PU sole or computer-aided footprint distribution (Fig. 48), known values can only be obtained from tests in wear,

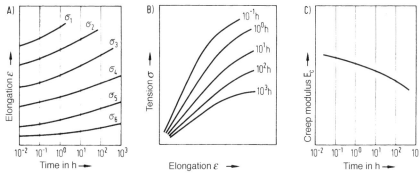

Fig. 47. Results and evaluation of creep tests
A) Creep curves; B) Isochronous stress-strain curves; C) Creep modulus time curves

with reliable documentation about stresses (user's weight, shoe size, frequency of wear, etc.). The same applies to measurement of the dynamic spring characteristic and the damping behaviour of car seats made of PU flexible moulded foam. Here too the test equipment comes quite close to the actual stress in the finished part under the load of a "standard bottom" (Fig. 49).

Practical simulation must be carried out under an actual load.

Ultimately, however, the consumer or user decides whether a product is a success or failure. In addition to the tests outlined here for determining mechanical properties, there are obviously very specific tests, for example with regard to ageing, resistance to weathering and dimensional stability in heat or cold and resistance to chemicals. The wear behaviour, for example of structural foam parts or PU elastomers, is defined by determining abrasion in the normal way. In the case of foams, it is important to determine water absorption, steam permeability and – particularly in the case of PU rigid foams – thermal conductivity. Airborne sound absorption is measured on elastic, open-cell PU foams. Other noise-reducing characteristics of polyurethanes (solid-borne noise, sound insulation) can also be established using DIN methods. The same applies to electrical and dielectric properties of polyurethanes.

In the end, the mass of data provided by short-term and sustained load tests, creep tests or tests on finished parts is not put aside, but becomes the systematic "raw material" for the construction of parts or the design

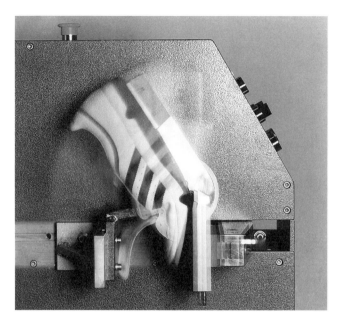

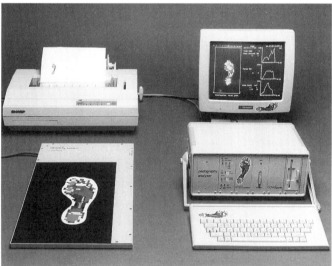

Fig. 48. Test equipment for flexural fatigue tests with the PU sole on the complete
shoe (top) and for measuring the distribution of the footprint (bottom) for
optimizing the PU sole for foot and substrate

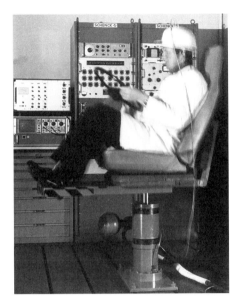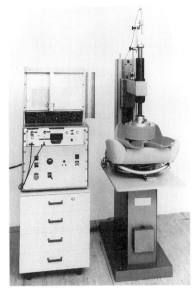

Fig. 49. Test equipment for car seats made of PU flexible moulded foam: "live" fatigue test in the pilot plant (left); measurement of the dynamic spring characteristic and damping behaviour on a standard bottom (right)

Table 19. FEM calculations for rheological, thermal and mechanical moulding design

Mould filling calculations for thin-walled mouldings made of solid polyurethane	– Establishment of a desirable gate location – Detection of entrapped air and its avoidance – Determination of the venting location – Determination of fibre orientations for RRIM – Determination of flow pressures and temperatures
Thermal calculations in the processing of solid PU systems	– Optimization of mould temperature control – Calculation of temperatures for estimating warpage
Mechanical calculations	– Deformations and stresses due to forces, moments and temperature changes – Dimensional stability: warpage due to cooling (isotropic and anisotropic)

of forms and moulds by the computer-aided finite element method (FEM).

The advantage of this method is that optimization can be carried out on the screen in an early phase of development and fine adjustments can be made to the prototype or production tool.

The lower run-in costs for production more than compensate for the higher development costs. Table 19 shows FEM calculations for PU technology.

5   How is polyurethane produced?

It has already been mentioned in Chapter 2 that polyurethanes are usually sold as intermediates, i.e. as liquid raw materials, rather than as a finished plastics material, for example TPU granules or fibres. As a result, firms described as PU "processors" do not "process" but manufacture polyurethanes – in contrast to thermoplastics processors who carry out physical forming. *PU manufacturers also carry out chemical and shaping operations.* This involves a special chemical reactor comprising containers equipped with stirrers, pumps, pipelines, valves, etc. In addition to these techniques which are typical of PU, finished polyurethanes can be processed and machined using a host of other techniques, including those found in the rubber industry. However, the following description will be restricted to the variation typical of polyurethanes.

5.1   General Technology

Liquid and sometimes also fused, dissolved and/or dispersed raw materials are reacted chemically during the production of polyurethanes. Heat is liberated in the process. There are two methods of reaction control, the components being allowed to react either simultaneously ("one-shot-process") or successively in two stages (prepolymer process) (cf. Section 6.2).

The construction and operation of PU plant, including transportation and storage of the raw materials, require official approval, for example under the Federal Pollution Control Act. Safety devices also have to be provided, for example protective metallic baths for storage tanks and other containers, overflow preventers, venting and temperature control systems as well as protective equipment for personnel (goggles, overalls, gloves, etc.). Fig. 50 is a block diagram of the construction and operation of PU plant. The main components A and B (polyisocyanate and polyol, see Section 6.3) are conveyed from storage tanks into working containers, brought to the prescribed temperature and fed by metering units to the mixing head. From there, the reaction mix is discharged onto a substrate or into a mould where it reacts fully. The raw materials for PU elastomers, which are frequently solid, first have to be melted, dewatered and degassed in the working container. In this case, one of the additives (components C to F in Fig. 50) is invariably a crosslinking agent or chain extender (see Section 6.3.3). By supplying additives, including gases, separately, the PU manufacturer is able to control formulations and therefore the properties of his product. The feed positions can be on the suction side and on the delivery side of the metering pump. Direct introduction into the mixing head is also feasible.

PU plant can be adapted to extreme processing conditions, if necessary:

- Processing of raw materials which are liquid at room temperature with viscosities of 5 to 20 000 mPas, including fused polyester polyols (see Section 6.3.2), at elevated temperature

- Metering and mixing of the raw materials in ratios of 1 : 100 to 1 : 1 for mouldings weighing a few g to 100 kg

- Adaptation of metering unit output to the reactivity of the system

- Discharge techniques for continuously operating plant and feed techniques for discontinuously operating plant

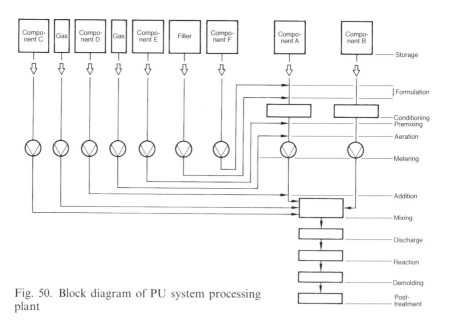

Fig. 50. Block diagram of PU system processing plant

• Processing of filler-containing components with granular, flaky or fibrous additives.

The development and use of suitable measuring, control, monitoring and data-processing systems allow a high level of automation in PU plant and machinery. Sensors and control elements provide assistance by measuring and checking the following process parameters:

• Temperature of raw material components

• Volumetric and mass flow rate; stoichiometric ratio of components

• Densities of components

• Mixing times

• Operating, injection and circulation pressures

• Gas content

• Mould temperatures

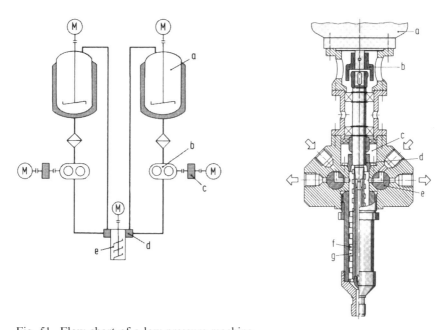

Fig. 51. Flow chart of a low-pressure machine
Left: a: working container; b: metering pump; c: gear; d: changeover devices;
e: mixing by stirrer mixing head for intermittent operation
Right: a: driving motor; b: clutch; c: lubricant supply; d: ring seal; e: changeover element; f: paddle stirrer; g: mixing chamber

So-called "live shift records" can also be obtained from graphic on-line displays. Finally, communication via modem/telephone/ISDN allows world-wide contact between machine suppliers and customers [16].

Slow gear pumps running at maximum working pressures of 40 bar are used for metering high-viscosity raw materials in *low-pressure machines.* Mixing is carried out by mechanical stirrers (Fig. 51).

This type of machine is also useful when producing mouldings weighing about 15 g with low total throughputs (2 l/min).

High pressure machines have various piston pumps for metering low-viscosity raw materials and convey the raw materials to an injection mixing head at working pressures of 100 to 300 bar. The reactants are mixed in a turbulent counterflow in the injection mixing head while utilizing their kinetic energy (Fig. 52 and 53).

Application of this mixing process is illustrated in Fig. 54 and in the photograph in Fig. 55.

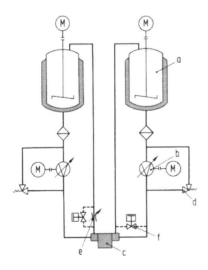

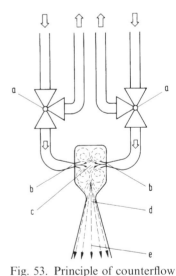

Fig. 52. Flow chart of a high-pressure machine (circulation system)
a: working container; b: metering unit; c: mixing head; d: safety valve; e: flow restrictor; f: low-pressure valve

Fig. 53. Principle of counterflow injection
a: changeover devices; b: injection nozzles; c: mixing chamber; d: flow restrictor; e: calming element

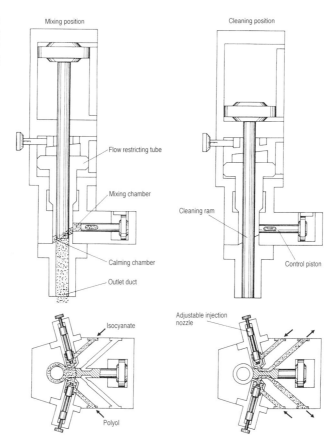

Fig. 54. Diagram of the MX mixing head (designed by Hennecke)

Fig. 55.
The attractive MX mixing head

Major advantages of the high-pressure process, apart from accurate metering and maintenance of the shot weight, are the possibility of processing "fast", i.e. highly reactive systems, minimal material loss and low environmental pollution, particularly when using self-cleaning injection mixers. The lower costs of a low-pressure machine do not offset these advantages. Table 20 gives an overview of processing techniques specific to polyurethanes.

With the *combination of high- and low-pressure techniques*, the components are injected under high pressure by piston pumps into a (low-pressure) stirrer-equipped mixing chamber. This method is adopted when more than two additives are to be metered separately in addition to the two main "ISO" and "POLY" reactants. The greater accuracy of the piston pumps is an advantage here.

The *closed-circuit principle* applied to the raw materials is common to both the low-pressure and the high-pressure techniques as the quality of the finished part is crucial.

Before each mixing and filling process the components circulate in the desired proportions and at the pressure required for injection. Forcibly controlled changeover devices cause a switch from circulation to injection in high-pressure injection mixers, all devices responsible for perfect, synchronized introduction of components into the mixing chamber being forcibly mechanically or hydraulically actuated. The reaction mix left in the mixing chamber is removed by cleaning pistons after the filling process or is blown away by compressed air.

Spraying machines can operate by the low-pressure or the high-pressure principle. A suitable jet is created by the additional air introduced in the first case and by the high mixing pressure in the second. Pneumatic driven single-cylinder piston pumps and tubing up to 100 m long are commonly used. The mixing head should be light and easy to handle. It is usually self-cleaning in design, the output of spraying machines being between 2 and 7 l/min.

5.2   Special Technology

A distinction is generally made between *continuously* operating plant for producing semi-finished products and *discontinuously* operating plant in

Table 20. PU systems and process engineering

PU systems/ stages of the process	1	2	3	4	5	6	7	8	9	10	11	12	13	14
Flexible foam														
Slabstock	X				X		X	X		X				
Moulded	X					X			X			X		X
Padding	X					X			X			X		X
Rigid foam														
Slabstock	X				X		X	X		X				
Cavity	X		X			X			X			X		X
Panels	X		X	X		X		X			X			
Pipes	X		X			X			X					
In-situ foam	X					X			X					
Structural foam hard + flexible	X		X	X		X			X			X	X	X
Elastomers														
Hot-cast		X					X		X			X		X
Cold-cast	X						X	X	X			X		X
Spray	X						X	X						
Casting resins	X						X		X			X		X
PU rubber		X							X			X		X
Thermoplastic	X	X					X	X						
	One-shot process	Prepolymer process	Premixing	Aeration	Feeding	High-pressure injection mixing	Low-pressure stirrer mixing	Continuous production	Discontinuous production	Slabstock plant	Laminator	Mould carrier	RIM plant	Demoulding

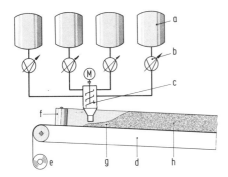

Fig. 56. Diagram of plant for the
continuous production of foam
slabstock
a: working container; b: metering unit;
c: stirrer-equipped mixing chamber;
d: conveyor belt; e: base paper; f: side
paper; g: reaction mix; h: fully reacted
foam

which open or closed moulds or cavities are filled with shots of reaction
mix.

Foam slabstock is usually produced *continuously*. This is illustrated in
principle by the *flexible slabstock foams* shown in Fig. 56. Rectangular
cross sections 2.20 m wide and 1.20 m high are common nowadays.
Outputs of 50 to 600 l/min of the reaction mix can be obtained with
foam block densities between 15 and 60 kg/m^3 and belt speeds of up to
10 m/min. This results in strip lengths of 30 to 100 m, depending on
formulation and curing time. Short (10 m) or long blocks of up to 120 m
leave the foaming plant and are left to cool in so-called reaction stores
for 12 hours until they acquire their final properties. There follow
conveyor belt systems, transverse cutters and trimmers as well as
conversion and fabricating machinery. Copy and contour cutting of
foam parts in unlimited shapes is made possible by CNC technology,
appropriate computer programs and water jet cutters. High-speed water
jet cutters (up to a maximum of 60 m/min) only use 0.11 to 1.18 l/min of
water, depending on pressure and nozzle size; the water pressure can be
varied from 500 to 3500 bar and the width of the sapphire or diamond
nozzles on the cutting head from 0.10 to 0.20 mm [17]. Furthermore,
sheets of optional length are produced in thicknesses of 0.5 to 30 mm on
carousel and long block slitting machines or roller peeling units. A
device is required for opening the partially closed cells when producing
high resilience foams. The moderate air-permeability of standard
flexible foams can be improved by wet (*Bauer process* with aqueous
alkali) or preferably dry chemical post-treatment. The foam in
pressurized vessels is subjected to an oxyhydrogen explosion by
electric ignition of an H_2/O_2 mixture, remaining cell membranes being

destroyed or melted onto the cell walls (*Chemotronics process*). The products obtained by this process are known as *reticulated foams. Rigid foam slabstock* is produced on similar plant and converted mainly into sheets which can be converted into building components by the *wrapping technique*, for example using sheet metal facings.

Rigid foam panels with or without facings are produced continuously on so-called *laminators*. In contrast to slabstock plant, in which large quantities of reaction mix have to be discharged, very thin layers are uniformly distributed on the substrate, for example by "rakes" (Fig. 57). In this case, the "wet part" of the "foaming plant" is preceded by winders for windable facings or feed units for rigid facings and profiling units for metallic facings which include kraft paper, aluminium, chipboard and plasterboard, steel from a coil or crimped sheet metal and coated glass fibre fabrics. It is followed by transverse cutters. Rigid foam panels are commercially available in widths of 0.5 to 1.30 m and thicknesses of 5 to 200 mm. Output in modern laminators can be as high as 20 m/min. Densities of 20 to 60 kg/m^3 necessitate outputs of 3 to 40 l/min, and hence laminators of between 12 and 30 m.

With *discontinuously* operating plant, the reaction mix is introduced intermittently into a cavity, for example PU rigid foam in refrigerators, or into a mould. The walls of lorry bodies (Fig. 22) are produced by the filling technique, the sheet steel hollow article being inserted into

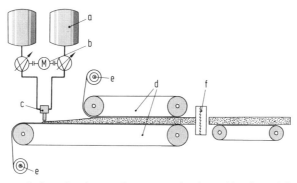

Fig. 57. Diagram of plant for the continuous production of laminated foam panels
a: working container; b: metering unit; c: mixing head; d: laminator; e: rollable facings; f: transverse cutter

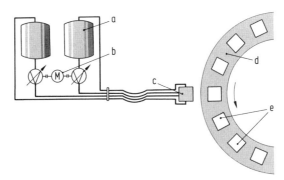

Fig. 58. Diagram of plant with moving moulds and a mixer head which moves in the feed direction
a: working container; b: metering unit; c: mixing head; d: turntable; e: moulds

press-like moulds and being filled with foam. Batch size, dimensions and contour of the moulding are crucial parameters in plant design. When producing small and miniature parts, the moulds are placed on carriers or smaller clamping units which advance continuously or intermittently. Turntables (Fig. 58) or oval belts are used.

The reaction mix can be poured into the open mould from at most two mixing heads. In the case of continuous mould carriers, the mixing head is moved along during the filling process. The reaction mix is introduced in spots or lines. In the case of mould carriers moved intermittently by the stop-and-go principle (plate feeders or turntables), the reaction mix can be introduced by robots to improve wetting of the mould surface. A 24-position turntable with two moulds per pallet can produce 600 parts/h. High-pressure machines are usually used in both systems. Examples include *flexible moulded foams* for car seats, which are produced as cold-cure or hot-cure foam depending on the raw material system used. In the first case, the mouldings pass through a device for opening the closed cells and in the second they pass through a heating tunnel where they cure completely prior to demoulding.

Stationary plant is more economical for large, complicated mouldings. If there are only a few, moulds at close intervals, they can be filled from the stationary machine via flexible lines and a mixing head or, alternatively, the plant can move past the moulds. Stationary plant has been successfully used for production of profile-like long mouldings, the moulds being star-shaped or semicircular. Examples include *flexible padding foams* which are produced by foaming behind facings (PVC or ABS sheets) in closed moulds. With highly reactive foam or solid systems, the filling rates are high and the curing times short, so it is not practical to move the mixing head away and then lock the mould. Therefore, the reaction mix for *structural foams* is injected into closed moulds, foaming and mould temperature being controlled so as to obtain mouldings with a cellular core and non-cellular outer zone. The transition from the interior to the exterior is continuous rather than abrupt (see Chapter 4, Fig. 39 and 40).

The term "reaction injection moulding" (RIM), or "R-RIM" in the case of reinforced materials, have been adopted for the methods of producing these highly reactive PU systems with very short moulding times (less than one second in some cases). High-pressure machines are normally used, a rigidly mounted, self-cleaning add-on mixing head being allocated to each mould with its clamping unit (Fig. 59).

The streams of component are distributed through closed circuit pipelines (see above). Glass in the form of ground short fibres (length = 140 to 240 μm) can be added to one of the components, usually the polyol (R-RIM) or, in the form of a glass mat, can be deposited in the mould and surrounded by the reaction mix in the second stage (S-RIM = structural RIM). With a more recent method (Fig. 60), cheaper long fibres (rovings) 10 cm long, typically 5 cm long, can be introduced directly into the mould with the reaction mix in one stage (LFI-PU = long fibre injection) [18].

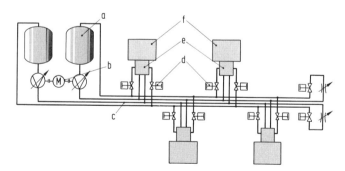

Fig. 59. Diagram of plant with stationary moulds, add-on mixing heads and multi-position metering (RIM plant)
a: working container; b: metering unit; c: closed circuit pipeline; d: emergency valves; e: add-on mixing heads; f: clamping units with moulds

Fig. 60. Direct introduction of
glass rovings and reaction mix
into the mould

There are significant differences between the RIM process and thermoplastic injection moulding: flow paths of up to 600 cm (with a wall
thickness of 2 mm) can be achieved with PU-RIM systems in comparison with 20 to 50 cm with thermoplastics. The internal pressures in
the moulds and the locking forces of 3 to 30 bar and 0.2 to 4 MN with
RIM are much lower than with injection moulding (70 to 300 bar and
25 to 200 MN).

Cavity foaming and moulding production invariably take place discontinuously. However, polyurethanes which are usually manufactured
continuously can also be produced discontinuously, for example in the
form of *flexible and rigid foam slabstock,* by a simple, inexpensive
method, adopted not just on account of its relatively low cost but
also on account of the throughput attainable (Fig. 61).

Within the experts jargon the so-called "box foam", or "golden bucket"
procedure, follows the "bottomless barrel" principle whereby a bottomless barrel is placed on the base of the box, is filled with hand-weighed
PU raw materials, the content is mixed and the barrel then pulled up.
The liquid reaction mix spreads over the base and expands; a "floating
lid" can be applied to avoid a bump of foam which has to be discarded.
After about 10 minutes, the finished foam slabstock can be demoulded.
This method can be economically employed for flexible foam produced
at rates of up to 800 t/a. Continuous plant is preferred for higher
throughputs.

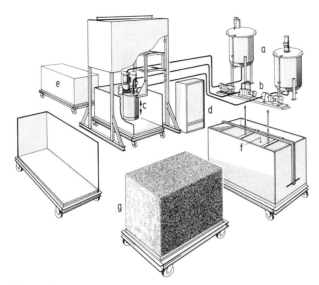

Fig. 61. Discontinuous production of flexible foam slabstock
a: raw material container; b: metering unit; c: batch mixer with lifting device;
d: switch cabinet; e: foam box; f: floating lid; g: finished foam slabstock

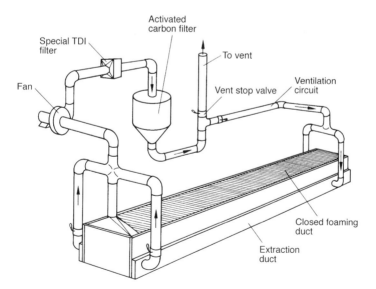

Fig. 62. Environment-friendly encapsulated plant for the production of flexible
foam slabstock ("E-Max. process")

To comply with environmental regulations, foam production by the discontinuous "long block process" takes place in completely encapsulated plant (Fig. 62). The ventilation system consists of combinations of filters.

In-situ foam based on PU rigid foam is produced as one- or two-component foam ("can foam"), either directly on site using transportable foaming machines or in small containers from pressure vessels.

5.3   Separate methods

The main elements of PU elastomer process engineering are prepolymer plant, hands, casting bench and machine. Hand processing is restricted to occasional large, heavy (100 kg) parts which would otherwise necessitate bulky machinery and equipment. Articles made of PU elastomers are usually produced by the prepolymer process. In specially designed plant (Fig. 63), the processor reacts polyols with an excess of diisocyanate to form a preadduct which is mixed with a crosslinking agent in the mixing head of a continuously operating casting machine in the second stage. The reaction mix (with a pot life of a few seconds or minutes) is rapidly introduced into heated moulds on the heated casting bench. Hot-cure *cast systems* are cured in ovens and cold-cure systems at room temperature.

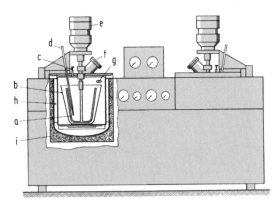

Fig. 63. Double reaction apparatus
a: anchor-type stirrer; b: casting pot; c: vent valve; d: thermometer; e: geared motor; f: filling nozzle; g: vacuum line to Wulfs' bottle; h: heating medium; i: insulation

Processing by the *one-shot process* is carried out under specific chemical conditions using low-pressure machines. Highly reactive systems for the reactive coating (with pot lives of 5 to 20 s) are processed (with compressed air) using automatically controlled casting or spraying units (roller coating) [19]. Airless processing using high-pressure machines is also feasible. *In-situ coating* in the open air (bridges, sports fields) is carried out using travelling metering, mixing and casting machines, and distribution of the reaction mass using applicator blades.

Thermoplastic polyurethanes (TPU) are produced discontinuously or continuously by the one-shot process. Once the components have been mixed in a reactor, the reaction mix is discharged onto a continuous belt. The TPU reacts fully and solidifies in the form of sheets which are chopped in a granulator and then processed into uniform granules (pellets, cylinders) in an extruder. Uniform granules are also obtained if the reaction is carried out in a twin-screw extruder. They can be processed using any commercial screw injection moulding machines equipped with a non-return valve. Extrusion is preferably carried out in a compression ratio of 1 : 2 to 1 : 3 using single-screw extruders with lengths of 20 to 30 D and comprising an evenly distributed three-zone screw. The materials are at temperatures between 180 and 250 °C depending on the type of PU, and the temperature of the moulds is generally kept at 20 to 40 °C by water. Pre-drying is not required for fresh and immediately processed granules, otherwise it is carried out at 100 to 110 °C for a period ranging from half an hour to a maximum of two hours. It is necessary in any case for clean, granulated scrap (sprues, etc.) which can be added in a proportion of up to 30 % to the fresh product for injection moulding. Extruder scrap should be converted separately or fed to the injection moulding equipment. Highly stressed finished parts attain their optimum properties after being conditioned for 15 to 20 h at 80 to 90 °C (up to 90 Shore A) or 110 to 120 °C (>90 Shore A), and untempered articles after 4 to 6 weeks' storage not below 20 °C.

The vulcanizable preliminary stage for *PU rubber* is produced by normal PU process engineering. Conversion, i.e. vulcanization, is carried out on normal rubber-manufacturing machinery. On account of the quite different chemical structure, however, a few peculiarities which will not be dealt with here should be noted. *PU lacquer raw*

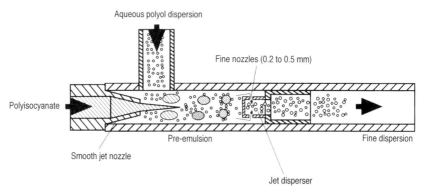

Fig. 64. Dispersion of aqueous PU lacquer systems by jets

materials are processed into *PU lacquers* and *PU coatings* by methods employed in the lacquer industry using special dispersing units, roll mills, fast-running dissolvers, attrition mills, etc. Jet dispersion is a novel method of mixing aqueous PU lacquer systems [20]. The hydrophobic polyisocyanate is pre-emulsified with the aqueous polyol dispersion using a smooth jet nozzle (Fig. 64) and is finely dispersed using a jet disperser: for obtaining PU powder coatings, the dry resin produced by PU processing methods is processed using extruders and is separated from the >100 m fraction by sieves after grinding the granules to a particle size of 40 to 60 m.

PU coatings are also produced using PU process engineering adapted to the specific chemistry. Fabrication, i.e. the coating of textiles, leather and paper, is carried out on machinery normally found in the relevant industry (direct, reverse roll coating). In reactive reverse-roll coating, the reaction mix is applied continuously to a base (mould) and is transferred onto the substrate (for example split leather) to cure. The plant resembles that shown in Fig. 57, so it will not be described in detail again.

However, the production of *poromeric PU synthetic leather* is a speciality of process engineering. Production of a "breathable" poromeric PU coating on a "textile" fabric is a complicated physico-chemical procedure. Without going into detail, the reader should simply glance at Fig. 65 and remember this next time he wears his (synthetic) leather coat.

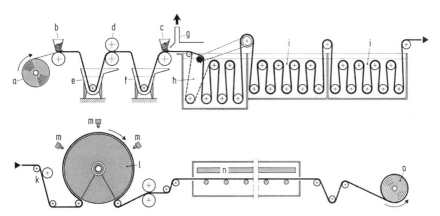

Fig. 65. Production of synthetic leather from breathable PU poromers: diagram of
the coagulation process
a: substrate; b: pre-coating; c: main coating; d: squeezing; e, f: small coagulation
baths; g: gelation air take-off; h: coagulation bath; i: washing baths; k: web control;
l: washing drum; m: spray nozzles; n: clamp frame dryer; o: winder

The machines used for processing fast-reacting PU components into
PU adhesives range from simple low-pressure machines with a working
pressure of 10 bar to high-pressure plant with working pressures of 200
to 400 bar, depending on reactivity and viscosity (see above). Otherwise,
simple apparatus equipped with stirrers is adequate for solvent-based
adhesives.

The use of PU chemicals in *binders* (PMDI), for example for *wood
shavings,* can reduce the pressing times on chipboard industry
machinery by 10 to 30%. An interesting variation from the PU
point of view is the processing of reactive PU systems into foundry
resins. A PU reactive material which only lasts for a few hours (2%
PU, 98% foundry sand) is supplied from storage hoppers directly to
the core shooting machine and is compressed into mouldings in core
boxes. These mouldings are cured within a few seconds at room temperature by blowing compressed air through an amine catalyst to form
ready-to-use cores (see Section 6.3.4) (Cold Box Process; Ashland).

PU fibres (elastanes) are produced continuously or discontinuously by
the prepolymer method. The spinning solution (22 to 36% of elastane)
is processed into filament yarns by the processes normally used for the

spinning and post-treatment of man-made fibres. The reactive spinning process is unique because PU and thread are formed simultaneously in the reactive spinning bath.

PU gels and *PU microcapsules* are produced in conventional chemical apparatus not specifically adapted for polyurethanes. High shearing forces are required for producing the very fine oil in water emulsion (3 to 10 μm) needed for microcapsules. Dispersers with frequencies right up to the ultrasonic range (Ultra-TurraxXREGX) are suitable for this purpose.

6 The chemistry must be right!

6.1 Chemical principles

As far as their chemical composition is concerned, polyurethanes will never be "pure" plastics like PVC, which is derived from vinyl chloride, or polyethylene, which is derived from ethylene, but will always be chemically and structurally mixed polymers. Quite often, the urethane group which is the godfather of this class of plastics (see Chapter 1) only represents a small part of the macromolecule. There are even "PU" products with no urethane groups! However, all variations are based on the principles of polyisocyanate chemistry. This term is more comprehensive than the historic name, "polyisocyanate polyaddition process", as distinct from polymerization and polycondensation, for the production of synthetic polymers. The actual properties of polyurethane are determined by features other than the urethane structure and this explains their versatility.

Isocyanates are characterized by the energyrich, i.e. very reactive isocyanate group, $-NCO$ which reacts exothermically not only with hydrogen-active compounds but, under the right conditions, with itself. The speed with which an NCO group reacts with its reactant depends not only on the NCO group but also on the structure of the molecular radical (see Section 6.4) to which it is bound. The most important hydrogen-active compounds contain $OH-$ or NH_2 functions. They are alcohols or amines. Water has an essential role as OH component in the production of foams. When isocyanate is used as a source of carbon, gaseous carbon dioxide CO_2 is formed and acts as a blowing agent. CO_2 is also produced when the NCO group reacts with organic acids – identified by the carboxyl group $-COOH$. Unconsumed or excess isocyanate in the reaction mix can enter secondary reactions with already formed primary reaction product. Other chemical structures which can permanently affect the properties of the finished polyurethane are also formed.

When isocyanates react with themselves, other structural elements are formed in the polymer and affect performance characteristics accordingly: in addition to the above-mentioned dimerization to carbodiimide, isocyanurate structures dramatically reduce the flammability of rigid

Table 21. Thermal resistance and stability to hydrolysis of polyurethane bonds

Thermal stability	Aliph.–NH–CO–NH–Aliph. n-Alkyl–NH–CO–O–n-Alkyl Aryl-NH–CO–O–n-Alkyl n-Alkyl-NH–CO–O-Aryl Aryl-NH–CO–O-Aryl –NH–CO–N–CO–O–	about 260 °C
		from 100
	R and –NH–CO–N–CO–NH– | R	(total from 150 to 160)
Resistance to hydrolysis (for example to 0.5 n-NaOH or 0.5 % HCl at 75 °C)	–NH–CO–O– –NH–CO–NH– –NH–CO–N(R)–CO–O and –NH–CO–N(R)–CO–NH–	≫–CO–O–

foams for the building industry (see Section 4.2). The reactivity of the – N=C=O-group is not restricted to these situations. It can also react with compounds which do not contain active hydrogen, for example with epoxides. The resultant products are used as embedding compounds and casting resins. The principles of polyisocyanate chemistry are shown as formulae of monofunctional examples and keywords in Fig. 66. The individual types of bonds have different stabilities to thermal and hydrolytic abuse, Table 21 showing some of their properties. They are used industrially as capped isocyanates in single-component PU systems, for the phosgene-free production of isocyanates and finally for the gylcolysis of PU scrap for obtaining recycled polyols.

6.2   Production and stoichiometry

Precise stoichiometric ratios, calculated using abbreviated formulae, must be adhered to when mixing the raw materials, which are generally liquid.

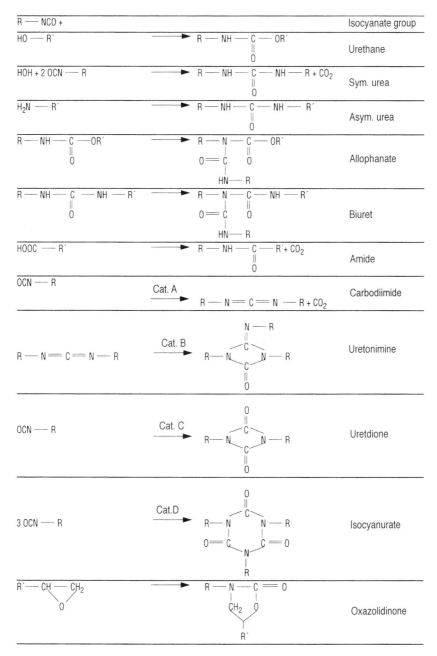

Fig. 66. Principles of polyisocyanate chemistry

Isocyanates are usually characterized by their NCO content in %, and sometimes by their equivalent weight, the term "functionality" also being used for valency:

$$\text{equivalent weight}_{isoc.} = \frac{\text{molecular weight}}{\text{valency}} = \frac{42 \cdot 100}{\% \ NCO}$$

Polyols are usually described by their hydroxyl value (OHV), sometimes also by % OH or alternatively by equivalent weight according to the following equations:

$$OHV = [\% \ OH] \cdot 33$$

$$\text{equivalent weight}_{polyol} = \frac{56 \cdot 1000}{OHV}$$

The stoichiometric quantities of *water, acids* (acid value AV), *primary and secondary amines* are calculated most easily using equivalent weights:

$$\text{equivalent weight } H_2O = 18:2 = 9$$

$$\text{equivalent weight }_{acid} = 5600:AV$$

$$\text{equivalent weight }_{prim.amine} = \text{molecular weight}:2$$

$$\text{equivalent weight }_{sec.amine} = \text{molecular weight}$$

The formulation is calculated by the equation:

$$1 \text{ g equivalent isocyanate} = 1 \text{ g equivalent polyol, etc.}$$

In practice, the quantity of isocyanate is almost always calculated for 100 parts by weight of H-active compound (for example polyol):

Example: How many g of isocyanate containing 48 % of NCO are required by 100 g of polyol with OHV 56 and 3 g of water?

$$\text{equivalent weight}_{isoc.} = (42 \cdot 100) : 48 = 87.5 \text{ g}$$

$$\text{equivalent weight}_{polyol} = (56 \cdot 1000) : 56 = 1000 \text{ g}$$

$$\text{equivalent weight}_{water} = 18 : 2 = 9 \text{ g}$$

quantity of isocyanate
for 3 g of water:
$$\frac{87.5}{9} = \frac{x}{3} \quad x = 29.2 \text{ g}$$

formulation: 100 g polyol and
                   3 g water are reacted with $8.75 + 29.2 =$
                   37.95 g isocyanate.

Other computations are often used for routine calculation of formulations with the same isocyanate. "Factors" ($F_{TDI} = 0.158$ and $F_{PMDI} = 0.242$) which provide the stoichiometric quantity for 100 g of polyol when multiplied by OHV are used in practice for the most common di- and polyisocyanates, TDI and PMDI. The *isocyanate index* is also an important parameter: it provides the percentage between actually used and the stoichiometrically calculated quantity of NCO:

$$\text{isocyanate index} = \frac{\text{actual quantity of isocyanate}}{\text{calculated quantity of isocyanate}} \cdot 100$$

By using the "free play" around the isocyanate index, PU manufacturers can influence the processability of the reaction mix or the properties of the polyurethane to a certain extent. An optimum range of isocyanate index is often recommended by raw material suppliers in the form of *starting formulations* (Table 22).

A manufacturer can obviously develop his own formulations. For economic as well as logistical and ecological reasons, he will preferably obtain a two-component system consisting of the polyisocyanate and the polyol component. The polyol component will contain all additives required for processing (see Section 6.3.4). This is also the case in the familiar "foam fungus" test for PU presentations (Fig. 6). During industrial processing, the foam manufacturer may have to add the blowing agent separately.

The time lapse between the mixing of components and the finished polyurethane might be queried at this stage. The timing of PU reactions is governed by the laws of physico-chemical kinetics. However, practitioners are not normally interested in these laws. Apart from the compulsory goggles and gloves, they are content with a stopwatch, ruler, beer mat and wooden skewer. A thermometer and a machete are also helpful for cutting open the foam to examine the cell structure. The progress of the complete PU reaction can be followed macroscopically using these simple aids. However, the research and development laboratory would have more sophisticated apparatus for taking kinetic measurements. Computer programs which not only calculate the

Table 22. Examples of polyurethane formulations

Flexible foam: Density: about 35 kg/m³	100 parts by wt. polyether (OH value = 45) 2.7 parts by wt. water 0.6 parts by wt. foam stabilizer 0.15 parts by wt. tert. amine catalyst 0.15 parts by wt. tin catalyst 35.8 parts by wt. TDI-80 *isocyanate index: 105*

Rigid foam:
Density: about 40 to 45 kg/m³
  (OHV = 356) (in the finished
  steel sandwich panel)
	100 parts by wt. polyol mixture incl. H_2O
	2.0 parts by wt. tert. amine catalyst
	16 parts by wt. HCFC R-141b
	147 parts by wt. PMDI
	isocyanate index: about 120

PU/PIR rigid foam:
Density: about 30 to 35 kg/m³
  (in the finished insulating panel)
	100 parts by wt. polyol mixture incl. H_2O
	  (OH value = 324)
	2 parts by wt. trim. cat. mix
	15 parts by wt. n-pentane
	170 parts by wt. PMDI
	isocyanate index: about 220

Hot-cast elastomer: Shore hardness: about 92 A	100 parts by wt. polyester polyol (OH value = 56) 50 parts by wt. MDI (pure)

left to react for about 1/2 hour at
  100 °C*)
add to room temperature
pour into hot mould (100 °C)
  immediately; can be demoulded
  after 10 to 20 min.; then
  condition for 24 h at 110 °C
	150 parts by wt. prepolymer (NCO about
	  10%)
	12.5 parts by wt. butanediol-1,4
	
	isocyanate index: about 108

* cf. footnote in Table 24

formulations but also show the actual PU reaction on the screen of a "virtual" reactor are also available [21]. The following processes and changes in properties are measured in the course of a PU foaming reaction:

- *Mixing time* (stirring time) is the period, beginning with 0, within which the reactants are mixed.

- *Cream time* is the time difference from time = 0 to a visible change in the still liquid mix (increase in viscosity and volume). The extremely fast rise in temperature of the exothermic PU reaction can be detected by the thermometer placed in the compound at this moment. Up to 150 °C can be attained in the laboratory formulation.

– *Curing time* (gel time) is the time difference from time = 0 to initial solidification (skewer! Keen cooks are familiar with this test).

– *Full rise time* (hold pressure time) is the time difference from time = 0 to attainment of the full foam height *h* (ruler!).

– *Tack-free time* is the time difference from time = 0 to that of a tack-free surface (test with a beer mat, not your finger).

A few seconds or minutes may have elapsed by now, depending on the reactivity of the PU system. Polymerization is not yet complete owing to the complexity of the concurrent PU reactions.

The foaming process is shown as $T = f(t)$ in Fig. 67 (top). It can be seen that these five reaction stages only apply to a fraction of the total reaction. This two-minute range is plotted schematically as $h = f(t)$. When producing mouldings, this can be followed by the

– *Mould dwell time* which is the period of time from $t = 0$ to the earliest possible demouldability ("green strength") of the moulding.

However, economic production is dependent on the

– *Cycle time* which includes preparation of the mould (opening, removal of moulding, cleaning, application of release agent, clamping).

Hours, days or weeks can elapse during the

– *Maturing time* from $t = 0$ until attainment of the optimum properties. It can be influenced by high temperatures (conditioning) and atmospheric humidity (cf. Sections 4.2 and 5.2).

Owing to the adjustable differences in reactivity both in isocyanates and in polyols and their formulations, PU production basically involves two methods of processing which have already been mentioned briefly in the foregoing chapters.

• The calculated total quantity of polyisocyanate is mixed with the calculated total quantity of polyol and additives in "one-shot" and is reacted: "*One-shot process*".

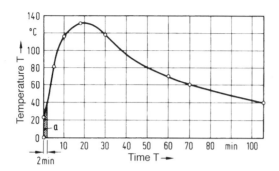

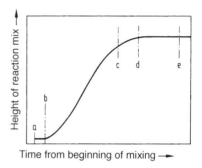

Fig. 67. The application-oriented
measured range (a; top) of the PU reaction
(in rigid foam) is narrow. The reaction
times can be measured in this short time
interval (bottom):
a: mixing time, b: cream time, c: curing
time, d: full rise time, e: tack-free time

- The total quantity of polyol (or chain extender) is reacted with a calculated excess (NCO/OH = 2) of polyisocyanate in a first stage. This product, which contains still reactive NCO groups, is reacted with crosslinking agents (or polyol) and the necessary additives to form the finished polyurethane in a second stage: "*prepolymer process*".

- If the ratio NCO/OH >2 is selected, the product is called *semiprepolymer*. It contains the prepolymer dissolved in excess diisocyanate which is then reacted by the above-mentioned process with the calculated residual quantities of polyol and additives to form the final PU.

- Pre-adducts in a ratio of NCO/OH ≫ 2 do not have a specific name in practice. Providing they are sufficiently stable in storage, they are sold as *modified isocyanates* – like (semi)prepolymers (see Section 6.3.1).

- If the reaction is carried out in a ratio of NCO/OH <1, OH-terminated (semi)prepolymers are obtained. These adducts are of limited importance in practice.

In comparison with the advantage of the "faster" one-shot process, the "slower" two-stage method via the prepolymer has the advantage of a more ordered structure of the PU polymer. Partial prepolymerization (NCO/OH $\gg$ 2) can also lead to liquefaction and therefore easier metering of a solid diisocyanate. The method actually used "*in situ*" depends on the type of polyurethane and the manufacturer's infrastructure, for example whether he is able to produce an NCO prepolymer himself. Raw material manufacturers normally supply ready-to-use prepolymers or *modified isocyanates* (see above) together with the "remaining" polyol and additives as a formulation and two-pack system. The PU manufacturer only performs the second stage of the prepolymer method and benefits from a quick one-shot process on site.

Monofunctional, including "CH_2 active", compounds are capable of reacting with isocyanates to form so-called "capped" isocyanates. These products behave inertly at room temperature in relation to polyols and only break down again at high temperatures into capping agent and isocyanate, which can then react with the polyol. Examples of capping agents include phenols, oximes, caprolactam and β-dicarbonyl compounds, for example malonic ester or acetic ester. Uretdione-modified diisocyanates also fall into this category. Capped isocyanates are used, for example, in one-pot systems (such as stoving lacquers).

The environment-conscious category of aqueous PU systems includes *PU ionomers* which contain ionic groups, for example $-NR_3^+$ or SO_3^- in the segmented polymer chain (see Section 6.4). These can also be obtained by reacting NCO prepolymers, for example with amino alcohols and by subsequent quaternization by methods employed in organic chemistry [22].

6.3   Raw materials and their origins

The raw material system for polyurethanes consists of three components: polyisocyanates (A), polyols (B) and additives (C). The names "A" and "B" components are not standardized and are used in different ways in different regions. The labels "ISO" for isocyanates and "POLY" for polyols are often also found on vessels.

The main sources of raw materials are petroleum, coal, salt, air and renewable natural materials (see Chapter 2). Fig. 68 shows the involvement of PU raw material chemistry in the overall complex of natural resources.

6.3.1 Di-/polyisocyanates

Isocyanates are produced on a large scale by phosgenation of amines. Chlorine circulated continuously under optimum conditions acts as the "chemical vehicle".

1. Rock salt electrolysis

$$2NaCl + 2H_2O - \text{electr. current} \rightarrow 2NaOH + H_2 + Cl_2$$

2. Phosgene production

$$2C + O_2 \rightarrow 2CO$$
$$2CO + 2Cl_2 \rightarrow 2COCl_2$$

3. Phosgenation

$$H_2N-R-NH_2 + 2COCl_2 \rightarrow OCN-R-NCO + 4HCl$$

4. Hydrochloric acid electrolysis

$$4HCl - \text{electr. current} \rightarrow 2H_2 + 2Cl_2$$

5. Hydrogenation of dinitro compounds

$$O_2N-R-NO_2 + 6H_2 \rightarrow H_2N-R-NH_2 + 4H_2O$$

Processes 2 and 3 are normally carried out in open-air industrial plant. The foreseeable risks of a phosgene leak are countered by an ingenious staggered safety system [6] (Fig. 69). Encapsulated equipment has been used for these two stages of the reaction since the end of 1992 (Fig. 70).

Between processes 2 and 4, the chlorine is circulated via hydrochloric acid from process 3. If the hydrochloric acid is sold, chlorine from process 1 is supplied at a later stage. Up to two-thirds of the hydrogen required to hydrogenate nitro compounds (process 5) into the amines for process 3 is evolved during electrolysis (processes 1 and 4). This is an example of circulation and interlinking which are familiar and desirable in chemistry. Because of the production process employed, the final isocyanates do not contain Cl_2 but only traces of chlorine in so-called "hydrolysable" or "acidic" form which is usually quantified in ppm in the manufacturer's product and safety data sheets.

Phosgene-free isocyanate synthesis has been attempted many times over the last 30 years but was unsuitable for the production of the "major"

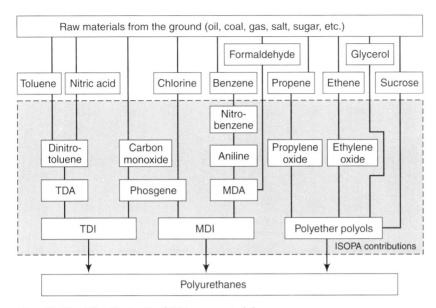

Fig. 68. The "family tree" of PU raw materials

Fig. 69. Fire drill at an isocyanate factory. Within fractions of a second, the plant is sprinkled with a water vapour wall from the exterior. In the event of a phosgene leak, ammonia is added to detoxify the phosgene

Fig. 70. Production of MDI in encapsulated plant. The phosgene-carrying parts are accommodated in steel containment 20 m in diameter and 35 m high

isocynates, MDI and TDI, for industrial, economic and ecological reasons. For one thing, secondary reactions and undesirable adjustments of equilibrium reduce yields: with annual global production of almost 2 million t MDI and over 1 million t TDI, a 1 % reduction in yield leads to generation of 20 000 t of additional waste.

In the case of special isocyanates, modified (bromine-consuming) Hofmann degradation is carried out if this is dictated by the economics governing the raw materials. For example, terephthalic acid which is available in the PET industry anyway can be converted by Hoffman degradation into *phenylene-1,4-diisocyanate (PHDI)* or in hydrogenated form into *cyclohexane-1,4-diisocyanate (CHDI)*. The ancient reaction between urea and alcohol (see Chapter 1), which gave

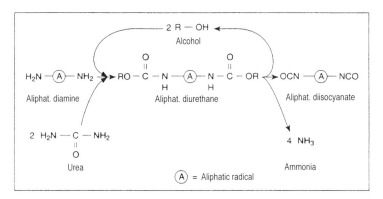

Fig. 71. Phosgene-free synthesis of aliphatic diisocyanates

urethanes their name, in the presence of a diamine and subsequent thermal decomposition of the diurethane yielded promising results (Fig. 71).

The alcohol is circulated. HDI has been produced by this process since 1994 (see Section 9.2). Another method has been discovered recently but is only suitable for minute quantities in the laboratory for reasons of cost. (Mono)isocyanates can be obtained in a yield of 97 % by reacting ditert.-butylcarbonate with amines [23].

Industrial polyurethane chemistry is entirely based on a few different types of base isocyanates which are listed in Fig. 72. Aromatic diisocyanates predominate, aliphatic diisocyanates, which are widely used in high-quality, lightfast PU lacquers appearing in smaller quantities. The most important *aromatic diisocyanates* are TDI and MDI. *Toluylene diisocyanate (TDI)* is derived from toluene. Nitration yields a mixture consisting of 80 % of 2,4- and 20 % of 2,6- or 65 % of 2,4- and 35 % of 2,6-dinitrotoluene (DNT), depending on the method of process control adopted. TDI-80, TDI-65 and TDI-100 types are obtained after hydrogenation (process 5) and subsequent phosgenation (process 3) (Fig. 73).

The numerical indices identify the content of more reactive 2,4-isomers in comparison with the less reactive 2,6-isomers (steric hindrance) in Table 23.

TDI-100 is the most active TDI. The least active TDI-"0", i.e. the pure

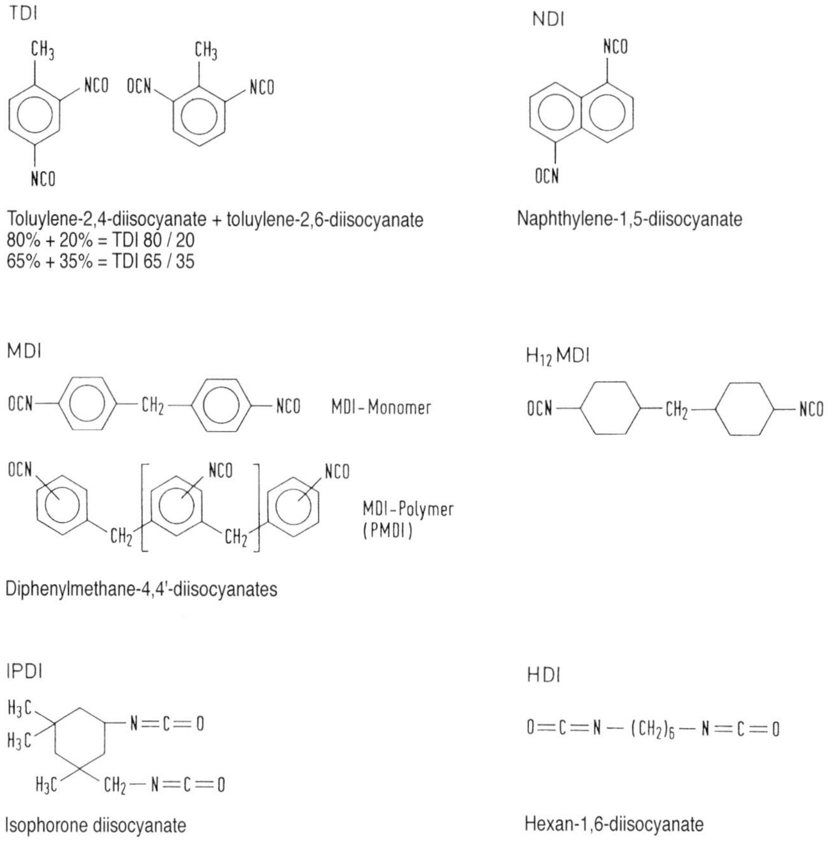

TDI

CH3   CH3

NCO   OCN   NCO

NCO

Toluylene-2,4-diisocyanate + toluylene-2,6-diisocyanate
80% + 20% = TDI 80 / 20
65% + 35% = TDI 65 / 35

NDI

NCO

OCN

Naphthylene-1,5-diisocyanate

MDI

OCN—⟨O⟩—CH₂—⟨O⟩—NCO   MDI-Monomer

OCN   NCO   NCO

        CH₂   CH₂

MDI-Polymer
(PMDI)

Diphenylmethane-4,4'-diisocyanates

H₁₂ MDI

OCN—⟨ ⟩—CH₂—⟨ ⟩—NCO

IPDI

H₃C
              N=C=O
H₃C

H₃C   CH₂—N=C=O

Isophorone diisocyanate

HDI

O=C=N — (CH₂)₆ — N=C=O

Hexan-1,6-diisocyanate

Fig. 72. The main industrial di- and polyisocyanates for PU chemistry

Table 23. Reaction rate constants of TDI-2,4 and TDI-2,6 in comparison with polyol and water

K ($1 \cdot mol^{-1} \cdot s^{-1}$)	Polyether polyol	Water
2,4-toluylene diisocyanate	$21 \cdot 10^{-4}$	$5.8 \cdot 10^{-4}$
2,6-toluylene diisocyanate	$7.4 \cdot 10^{-4}$	$4.2 \cdot 10^{-4}$

2,6-isomer, is not used industrially as it is not easily available, even though its symmetry suggests interesting PU properties [24]. Industrial TDI products are distilled, pure liquids which are clear when fresh and move readily. TDI-80 is the most important diisocyanate for PU *flexible foams*. In terms of quantity, TDI was outstripped by

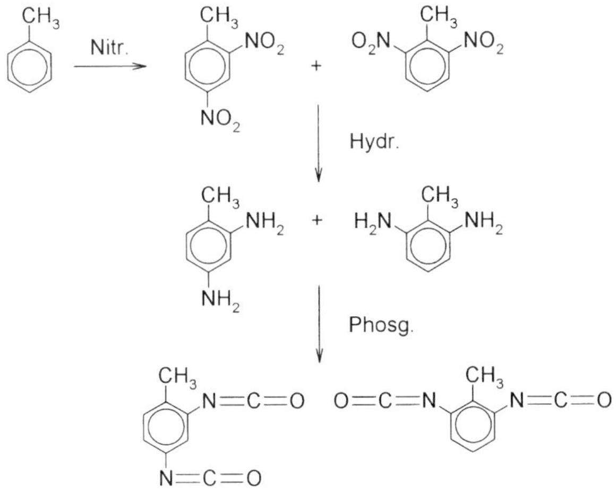

Fig. 73. Synthesis of
toluylene diisocyanate

diphenylmethane-4,4'-diisocyanate (MDI) a long time ago. The abbreviation "MDI" is derived from the old name methylene diphenyl diisocyanate. The diamine MDA which is to be phosgenated by process 3 (see above) can be obtained from benzene via nitrobenzene and aniline and by condensing it with formaldehyde (Fig. 74).

A wider mixture of isomers and homologues is produced by this process than by the nitration of toluylene for obtaining TDI. This can be seen from Fig. 75. The range of isomers is extended beyond the position isomerism of the NCO groups on the "binuclear" products by condensates with 3, 4 and more aromatic nuclei. The diphenylmethane-4,4'-diisocyanate – the "actual" MDI – is distilled in part from the crude

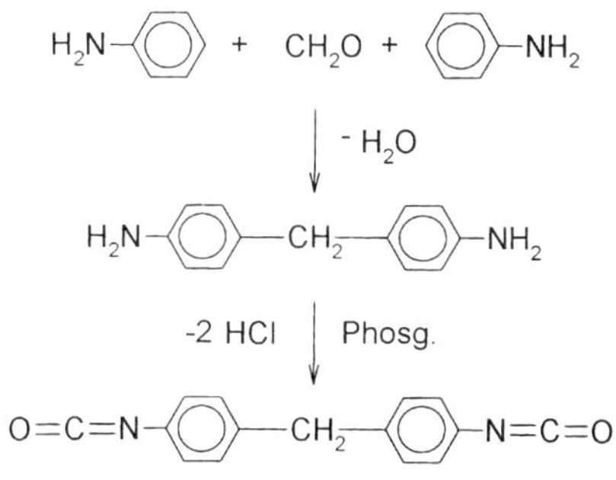

Fig. 74. Synthesis of diphenylmethane-
4,4'-diisocyanate (MDI)

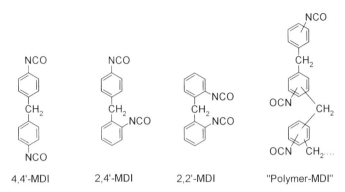

Fig. 75. MDI isomers and polymeric MDI

phosgenation product. It contains small quantities (<2 %) of the 2,4'-isomer and is used predominantly for *PU elastomers*. The majority is still a mixture of binuclear and higher nuclear individuals known as *polymeric MDI (PMDI)*. It has functionality >2 and is a standard low-viscosity (a few 10^2 mPa s), light – formerly dark brown – product, depending on the process control employed. It is the most important polyisocyanate for producing *PU rigid foams*. Some 2,4'-MDI-rich types are also suitable for PU flexible moulded foams.

TDI and (P)MDI are also the base isocyanates for an extensive range of *modified isocyanates* (see Section 6.2). Raw material manufacturers make use of the almost inexhaustible possibilities for reactions between H-active compounds and diisocyanates, on the one hand, and secondary reactions sketched in Fig. 72, on the other hand. Modified isocyanates are widely used for special applications over the entire range of polyurethanes.

Special products include *naphthylene-1,5-diisocyanate (NDI) for top-quality PU cast elastomers* and higher functional *tri- and tetraisocyanates* based on triphenylmethane *for adhesives*.

Aliphatic diisocyanates can be used in the paints and coatings field owing to their lightfastness. The diamine for phosgenation (process 3) forming the basis for *hexane-1,6-diisocyanate (HDI)* can be obtained by hydrogenation of adiponitrile which, in turn, can be obtained by various methods of industrial synthesis. Reference has already been made to the phosgene-free conversion of HDA to HDI (see above).

HDI is used exclusively as modified (biuretized) higher molecular weight polyisocyanate. *Isophorone diisocyanate (IPDI)*, whose intermediate IPDA originates from acetone chemistry, is also used as modified isocyanate for lacquer.

6.3.2 Polyols, polyamines

The products listed here are the most important reactants for isocyanates. There are more commercially available types of polyols than the number of isocyanates to the power of ten. The enormous range of polyurethanes can be decisively influenced by applying knowledge and experience when combining various polyols with polyamines and/or crosslinking agents (Fig. 76).

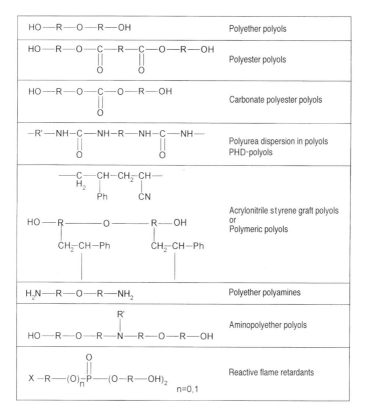

Fig. 76. Main industrial reactants for di- and polyisocyanates in PU chemistry

Polyether polyols are the most commonly used products. They form the "backbone" of polyurethanes. A wide range of long- and short-chained polyether polyols with 2 to 8 OH groups per molecule (functionality) can be obtained by a base-catalyzed (KOH) reaction between di- and polyhydric "starting" alcohols with epoxides (propylene and/or ethylene oxide) (Fig. 77). Whereas polyethers with secondary OH groups are obtained with propylene oxide ($R'=CH_3$), the reaction with ethylene oxide ($R'=H$) leads to more highly reactive primary OH groups. Polyether activity can be adjusted by carefully "playing" with epoxide. Examples of starting alcohols with a synthetic base include glycols, trimethylolpropane, pentaerythritol and glycerol; glycerol is also available from renewable natural substances such as α-methylglycoside, sorbitol, sugar, hydrolysed starches, etc. Polyethers with a particularly low mono-ol content and a close molecular weight distribution can be obtained by using a special catalyst (zinc hexacyano cobaltate). They are particularly suitable for prepolymers for elastomer production [25]. Diamines (for example ethylene diamine) or alkanol amines are also reacted with epoxides to form amino polyether polyols. They must be clearly distinguished from polyether amines (see below). Polyether polyols are substantially stable to hydrolysis; they have to be stabilized

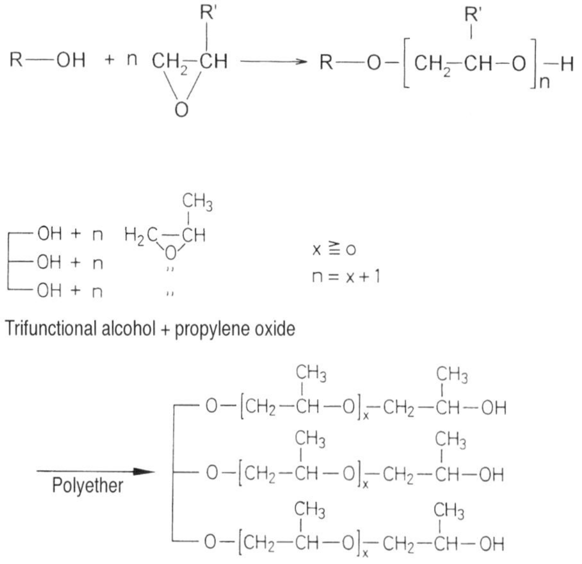

Trifunctional alcohol + propylene oxide

Fig. 77. Polyether production, theoretical (top) and using a trihydric alcohol (bottom)

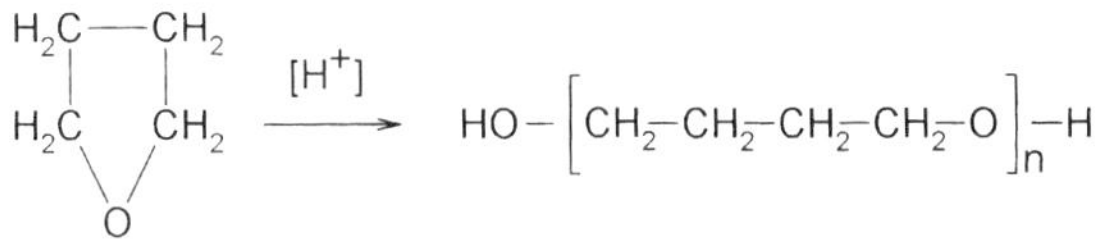

Fig. 78. Production of C_4 ether by polymerization of tetrahydrofuran

against photooxidation by using additives. So-called *C_4-polyethers* are less sensitive to photooxidation [26]. Strictly linear polyether diols are obtained from tetrahydrofurane (THF) by ring-opening polymerization (Fig. 78). THF can be obtained both synthetically and from renewable natural substances.

Polyester polyols are used to a much lesser extent that polyether polyols. They are more expensive to produce (Fig. 79) and much more viscous than polyethers with comparable chain lengths. On the other hand, they are far less sensitive to photooxidation but susceptible to hydrolysis. Common starting materials include industrially available adipic acid along with teraphthalic acid and low molecular weight diols such as glycol, butanediol-1,4, glycerol, etc. The proportion of byproducts (for example cyclic esters) capable of evaporating can be drastically reduced by modern methods. The polyester flexible foams used for interior trims for cars are "low-fogging" and reduce misting of windscreens (<1 mg of condensate according to DIN 75201/limit for the automotive industry). *Carbonate ester polyols* (Fig. 76) which can be obtained by ester interchange of diphenylcarbonate with diols (for example hexanediol) are special polyesters. The *poly-ε-caprolactone polyols* obtainable by Ti-catalysed ring-opening polymerization of ε-caprolactone with diols or triols are another special type (Fig. 80). Both types are more resistant to hydrolysis than conventional polyester polyols. One reason for this is that the chain between the ester bonds is one or two carbon atoms

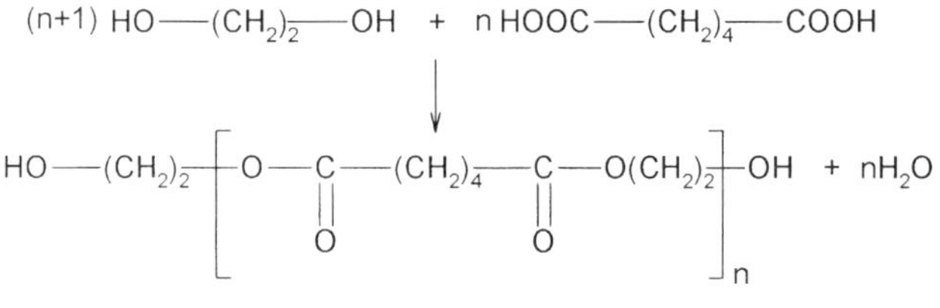

Fig. 79. Theoretical polyester production

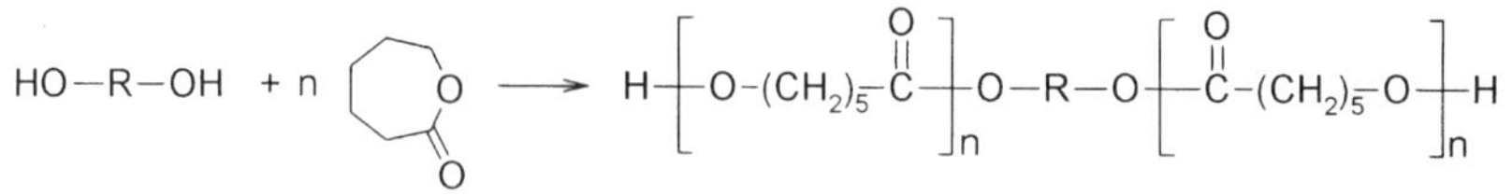

$$HO-R-OH \; + \; n \; \bigg(\! \! \overset{O}{\underset{O}{\big)}} \! \! \longrightarrow \; H \! \left[O\text{-}(CH_2)_5\text{-}\overset{O}{\overset{\|}{C}} \right]_n \! \! O\text{-}R\text{-}O \! \left[\overset{O}{\overset{\|}{C}}\text{-}(CH_2)_5\text{-}O \right]_n \! \! H$$

Fig. 80. Obtaining of polycaprolactone polyol by glycol-initiated polymerization of ε-caprolactone

(hexane diol-carbonate) longer than in adipic acid. The undisputed advantage of polyester polyols lies in their contribution to the high strength of the final polyurethane. This is due to their ability to form hydrogen bridge bonds, for which polyethers are not so well suited. The special ester polyols based on carbonate and ε-caprolactone are used mainly for high-quality PU elastomers.

Diols based on polybutadiene [27], *ethylene butylene copolymers* [28], and *hydrogenated polyisoprene* [29] have an oxygen-free polymer chain (Fig. 81 to 83).

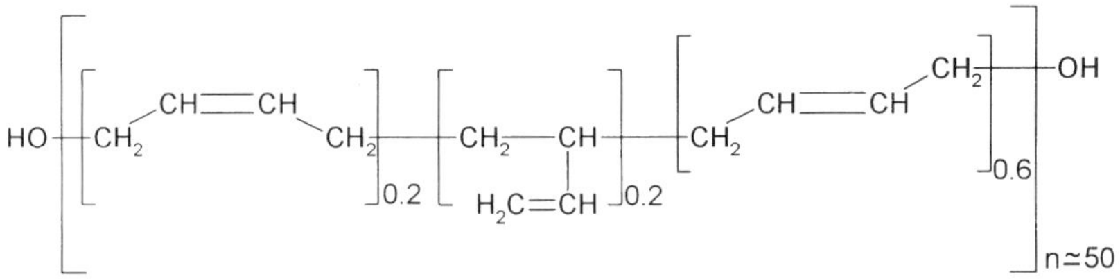

Fig. 81. Polybutadiene polyol [24]

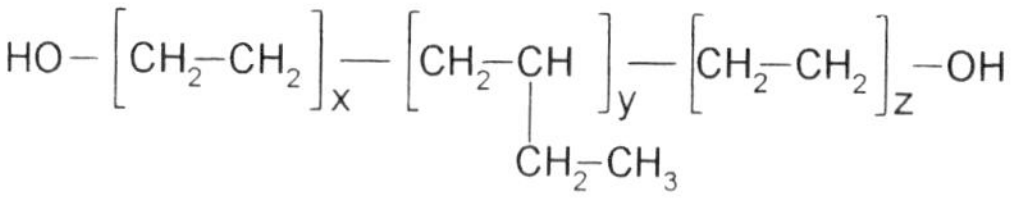

$$HO - \left[CH_2\text{-}CH_2 \right]_x - \left[\underset{\underset{CH_2\text{-}CH_3}{|}}{CH_2\text{-}CH} \right]_y - \left[CH_2\text{-}CH_2 \right]_z - OH$$

Fig. 82. Poly(ethylene/butylene) diol [28]

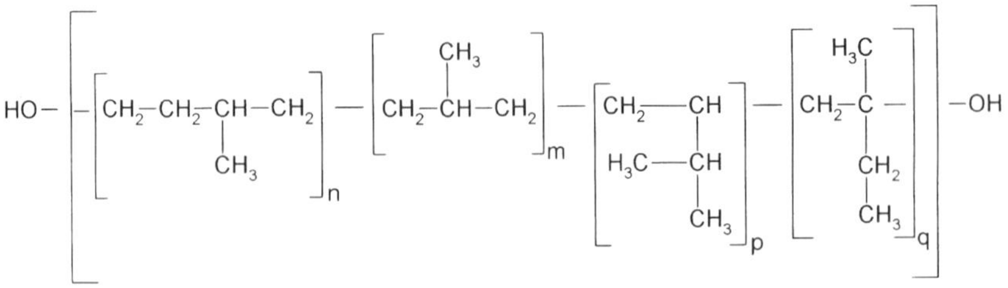

Fig. 83. Hydrogenated polyisoprene diol [29]

The main fields of application are PU elastomers and other non-cellular polyurethanes. A specialist can easily see that polybutadiene diols have to be protected from O_2/O_3-ageing by stabilizers; on the other hand polyethylene/butylene diols are resistant to ultraviolet light.

Organically filled polyols increase the indentation resistance and elasticity of flexible foams with relatively low densities. The filler polyols are also used in semi-rigid structural foams and PU elastomers. They are known, on the one hand, as graft, SAN or polymeric polyols and, on the other hand, as polyurea dispersion polyols. These are milky white stable dispersions of styrene/acrylonitrile and more recently only of styrene polymers [30] in the first case and of polyureas in the second case. In a special variation, the "foamer" produces a PU dispersion as an OH semi-prepolymer with a limited shelf life from polyether, possibly OH crosslinking agent and TDI ("PIPA").

(Poly)amines are highly reactive toward isocyanates. They are higher or lower molecular weight compounds with two or more NH_2 groups. The former include polyether amines while the latter pass over to the crosslinking agents/chain extenders. The urea segments incorporated into the PU framework via the NH_2 group give the polymer rigidity and improved high and low temperature resistance. The polyaddition reaction rate can also be increased by using amino compounds.

Tri-OH-functional castor oil (see Section 6.5) and tall oil formed during the production of cellulose are also used directly as *renewable raw materials* in formulations.

6.3.3 Crosslinking agent/chain extenders

The two terms are often erroneously used as synonyms. This may be due to the fact that processors are more familiar with "crosslinking" than chain extension. Crosslinking agents and chain extenders are both low molecular weight diols or triols and diamines so the same chemical can even perform both roles, in other words the process rather than the specific chemical determines whether a diol is a crosslinking agent or chain extender. This is illustrated by the following examples (Fig. 84 and 85).

It can be concluded that: although a chain extender leads to a "longer" but still functional reactive intermediate product (for example

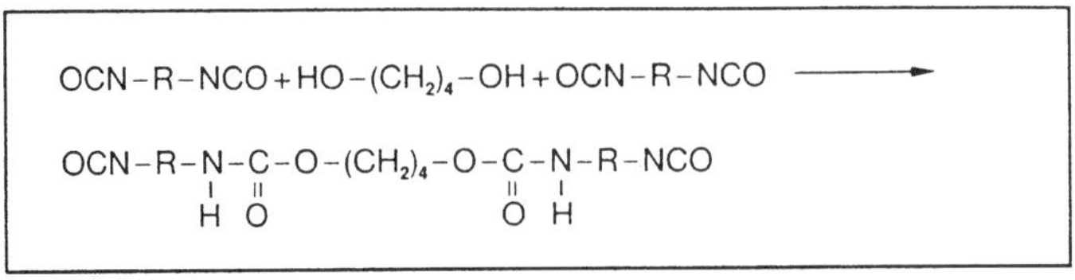

OCN−R−NCO + HO−(CH₂)₄−OH + OCN−R−NCO ⟶

OCN−R−N−C−O−(CH₂)₄−O−C−N−R−NCO
　　　　|　||　　　　　　　　||　|
　　　　H　O　　　　　　　　O　H

Fig. 84. A diisocyanate is "extended" with butanediol

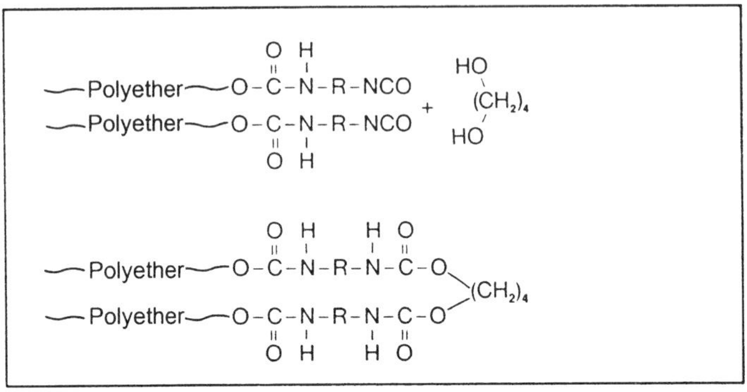

```
              O  H
              ||  |                 HO
~Polyether~O−C−N−R−NCO              \
                            +        (CH₂)₄
~Polyether~O−C−N−R−NCO              /
              ||  |                 HO
              O  H


              O  H    H  O
              ||  |    |  ||
~Polyether~O−C−N−R−N−C−O
                                \
                                 (CH₂)₄
~Polyether~O−C−N−R−N−C−O
              ||  |    |  ||    /
              O  H    H  O
```

Fig. 85. An NCO prepolymer is "crosslinked" by butanediol

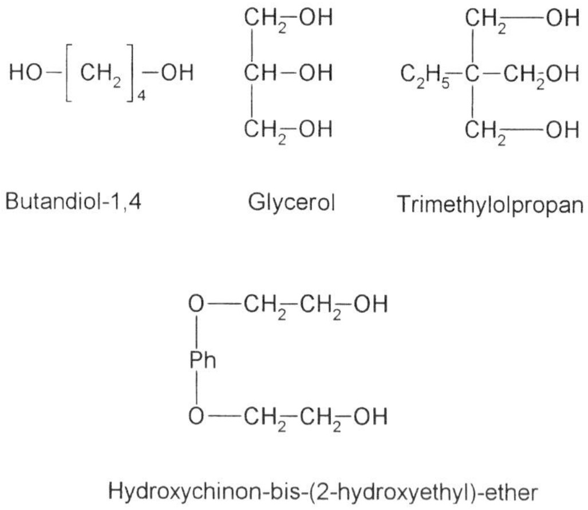

```
                    CH₂−OH          CH₂——OH
                    |               |
HO−[ CH₂ ]₄−OH      CH−OH     C₂H₅−C−CH₂OH
                    |               |
                    CH₂−OH          CH₂——OH

Butandiol-1,4       Glycerol        Trimethylolpropan
```

```
O——CH₂−CH₂−OH
|
Ph
|
O——CH₂−CH₂−OH

Hydroxychinon-bis-(2-hydroxyethyl)-ether
```

Fig. 86. Examples of OH crosslinking agents/chain extenders

urethane-modified diisocyanate, see above), a crosslinking agent leads to the "crosslinked" final product, polyurethane.

Di- and polyhydric alcohols such as butanediol, glycerol or trimethylol propane are used as OH crosslinking agent/chain extender (Fig. 86).

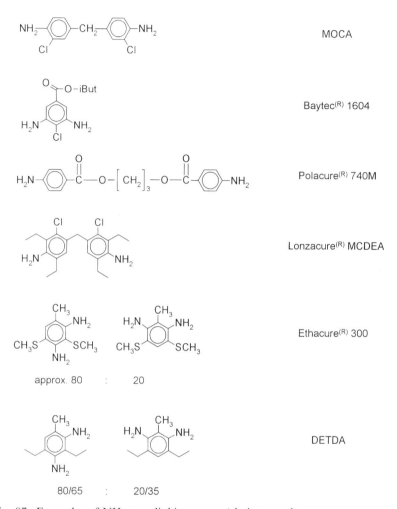

Fig. 87. Examples of NH$_2$ crosslinking agents/chain extenders

Aromatic diamines, whose NH$_2$ groups react much more slowly with NCO due to the steric hindrance caused by neighbouring alkyl groups or Cl atoms, are used almost exclusively as NH$_2$ crosslinking agents/ chain extenders (Fig. 87).

The network density of the PU polymer can be controlled using crosslinking agents/chain extenders and its properties thereby influenced. Urethane structures are obtained with OH crosslinking agents/chain extenders and urea structures with NH_2 crosslinking agents/chain extenders.

A different crosslinking principle is involved during the cross linking of *PU rubber*. On the one hand, the PU rubber can be crosslinked as OH prepolymer with a capped, usually uretdione, diisocyanate. On the other hand, vulcanization with sulphur is permitted by introduction of double bonds into the prepolymer. Finally, a polyester/MDI polyurethane can also be crosslinked radically on the CH_2 groups of the MDI by using cumyl peroxide.

6.3.4 Additives

Apart from the basic raw materials, polyisocyanate and polyol, additives or auxiliaries are also required for producing polyurethanes. The PU manufacturer only comes into direct contact with these chemicals if he is developing his own formulations. Most of the necessary additives are contained in the ready-to-process two-component systems formulated by raw material suppliers (Table 24):

Catalysts accelerate the reaction, tertiary amines and/or organo tin compounds usually being used. Table 25 lists the types of chemical structure while Fig. 88 shows some commonly used catalysts.

Table 24. Additives for the production of polyurethanes

Additives	Required
Catalysts	always*
Crosslinking agents/chain extenders	frequently
Surfactants	usually
Blowing agents	for foams
Flame retardants	as necessary
Fillers	as necessary
Antioxidants	frequently/usually
Release agents	for mouldings
Colourants	as necessary
Special additives (e.g. biocides, antistatics)	as necessary

* unless a supply of heat can replace the catalyst (cf. Table 22)

Table 25. Structural features of industrially important PU catalysts

Structure	Description
$R_1\diagdown$ $\quad$ R_5 $\quad$ R_3 $\quad\quad$ $R_1\diagdown$ N $\diagup R_3$ $\quad$ N$-R_n-[\overset{\|}{N_1}]-R_m-N\diagup R_4$, $\quad$ R_2 $R_2\diagup$	n = 2 to 6 and higher m = 2 to 3 l = 0 (m = o) to 4 (cyclo-)aliphatic tert. amines
$R_1\diagdown$ $\quad\quad$ $\diagup R_3$ $\quad$ N$-R-O-R-N$ $R_2\diagup$ $\quad\quad$ $\diagdown R_4$	(cyclo-)aliphatic amino ethers optionally incl. heteroatoms
$R_1\diagdown$ $\quad$ $\overset{R_4}{\overset{\|}{C}}$ $\quad$ N$-C=N-R_3$ $R_2\diagup$	(cyclo-)alipatic amidines; heterocycles
$\quad\quad$ O $\quad\quad$ $\|$ $R_1\diagdown$ $\diagup O-C-R_2$ $\quad$ Sn $R_1\diagup$ $\diagdown O-C-R_2$ $\quad\quad$ $\|$ $\quad\quad$ O	R_1 = alkyl (frequently C_4); halogen (rare) R_2 = alkyl (frequently C_5-C_{11}), ester (mercaptides, thioester) of divalent (R_1 = 0) and tetravalent tin
R$-$COOMe	R = alkyl Me = alkali metal (frequent)

Tertiary amines

Triethylamine	$N(C_2H_5)_3$
[R]Desmorapid DB	(phenyl)$\diagdown N(CH_3)_2$
[R]DABCO	(structure)
Catalyst AI	$H_3C\diagdown N\diagdown\diagup O\diagup N\diagup CH_3$ ... $H_3C\diagup$ $\quad\quad\quad\quad\quad$ $\diagdown CH_3$

Organo-tin compounds

Dibutyl-tin dilaurate (DBTL)	$H_9C_4\diagdown$ $\quad$ $O-\overset{O}{\overset{\|}{C}}-C_{11}H_{32}$ $\quad\quad$ Sn $H_9C_4\diagup$ $\quad$ $O-\underset{O}{\underset{\|}{C}}-C_{11}H_{32}$
Tin dioctoate (more precisely: tin-2-ethylhexoate)	$Sn(O-\underset{O}{\underset{\|}{C}}-\overset{C_2H_5}{\underset{H}{\overset{\|}{C}}}-(CH_2)_3-CH_3)_2$

Fig. 88. Commercial catalysts (selection)

Surfactants, for example emulsifiers, improve the miscibility of the reactants, polyisocyanate/polyol/water, which are actually "incompatible" and, when combined with catalysts, contribute to a uniform PU reaction. Special organo-silicon compounds are used as *foam stabilizers* and/or cell regulators during foam production (Fig. 89). They stabilize the rising foam until it cures and also regulate the open and closed cell character and pore size of the foams.

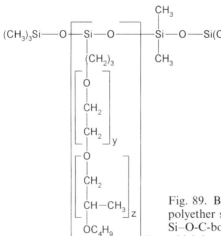

Fig. 89. Basic structure of commercial polyether siloxanes as foam stabilizers: with a Si–O–C-bond at the top and with a Si–C-bond which is more stable to hydrolysis at the bottom

Blowing agents are used for producing foams from the liquid, viscous reaction mix. A distinction is made between *chemical* and *physical* blowing methods. The former is based on the reaction between isocyanate and water (Fig. 66) and evolves gaseous carbon dioxide (CO_2) as blowing gas. Another chemical blowing method is based on the decomposition of specific formiates into carbon monoxide CO. It is less important than the above-mentioned CO_2 reaction. With the *physical* method, the addition of low-boiling point liquids causes the exothermically reacting mixture to expand owing to evaporation of the blowing agent. H(C)FCs and/or hydrocarbons (pentane, cyclopentane) which are less harmful to the ozone layer are used nowadays instead of the former CFCs (*so-called "first generation" blowing agents*) in order to protect the climate [31]. Formulation chemistry, in particular catalysis, also has to be adapted to this [32]. Research and development into industrial, economic and ecological aspects, including complex toxicological tests, to find out which blowing agent is the best substitute for CFCs has not yet been completed [33]. Decision-making is not simplified by the frequent changes to the deadline for CFC elimination. At the Vienna Conference (1995) attended by signatories of the Montreal Agreement, (*"second generation"*) HCFC's with a much lower ODP (ozone depleting potential) were designated as a transitional solution, at least in industrial countries. R-141b [34] is particularly affected; it is currently used throughout the world (Fig. 90).

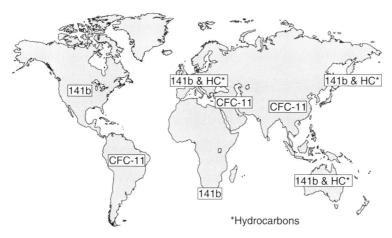

Fig. 90. Geographic distribution of blowing agents used for PU foams in 1996

The efforts made in pursuance of the Montreal Protocol have borne fruit: ODS (ozone depleting substances) values in the lower atmosphere fell in 1995. Recovery, i.e. closure of the hole in the ozone above the Antarctic, is expected within the next decade, providing that industrial and economic efforts to replace CFC's are maintained. This applies in particular to regions where R-11 is still used as a blowing agent even though scenarios for elimination have been established there – but there is no money: according to estimates, for example, Indian firms will have to find about 2 billion US dollars in order to comply with the Montreal deadline for elimination (2010 for India) [37].

240 000 t of CFCs are still produced globally [38]. *"Third generation"* HFCs are blowing agents with an ODP = 0 and low greenhouse supplementing effect. These include R-134a, formerly developed as a refrigerant whose low boiling point ($-26\,°C$) and modest solubility in polyol formulations are a drawback in PU technology. On the other hand, research and development [39] is being carried out into the HFCs

R-245 fa: pentafluoropropane $CF_3-CH_2-CHF_2$ [35]
R-365 mfc: pentafluorobutane $CF_3-CH_2-CF_2-CH_3$
and   R-356 mff: hexafluorobutane $CF_3-CH_2-CH_2-CF_3$

Table 26 shows a few cell gases, again in comparison with CFCs.

Processes which employ the almost ideal high-temperature/low-temperature insulating capacity of a vacuum make use of the thermos flask principle (Dewar flask). Where there is "nothing", "nothing", that is no heat, can be conveyed. Hence, *open-cell* PU rigid foam sheets produced continuously on laminators can be welded into film in a gas tight manner under vacuum in the presence of a getter layer and can be incorporated in refrigerators (Fig. 91) [40].

Liquid CO_2 under pressure is also used, for example, for PU flexible foam production. As it can be obtained inexpensively from the air and other natural sources, it is "greenhouse neutral" and easy to handle in the appropriate state (Fig. 92). As well as in bottles, it is sold at $-28\,°C$ at 15 bar pressure in cold-insulated road tankers and can be stored at $-20\,°C$ and 18 bar in tanks hired from the manufacturer. As CO_{2liq} is transported from storage tank to mixing head in liquid form, the stream of substance must be absolutely gas-tight. Adaptations or modifications to existing or new foaming plant are restricted to a mixing head

Table 26. Properties of various blowing agents for PU foams

Cell gas	Formula	Boiling point °C	Thermal conductivity (W/K·m)·10^{-3}	Flammability	ODP[b]	GWP[b]	Shelf life[b] Years
CFC 11	CCl_3F	+24	8.0/30°C	—	1	1	60
CFC 12	CCl_2F_2	−30	9.9/30°C	—	0.92 to 1.0	2.8 to 3.4	120
HCFC 123	$CHCl_2$–CF_3	+28	9.0/21°C		0.013 to 0.019	0.017 to 0.02	1.6
HCFC 141b	CH_3–CCl_2F	+32	7.9/21°C	flammable	0.066 to 0.092*	0.087 to 0.097	7.8
HCFC 22	$CHClF_2$	−41	11.0/30°C	—	0.042 to 0.057	0.34 to 0.37	15.3
HCFC 142b	CH_3–$CClF_2$	−10	9.4/21°C	flammable	0.053 to 0.059	0.34 to 0.39	19.1
HFC 134a	CH_2F–CF_3	−26.5	14.1/60°C		0	0.5 to 0.29	15.5
HFC 152a	CH_3–CHF_2	−25	9.4/25°C	flammable	0	0.026 to 0.033	1.7
n-pentane (isopentane)	C_5H_{12}	+36 (+28)	13/30°C	flammable	0	0.00044 (0.00037)	4.8 (7)[c] days
Air	N_2/O_2	—	29/30°C	—	0	—	—
Carbon dioxide	CO_2	sublimes −78.5	17/30°C	—	0	about 0.00025[d]	about 120[d]

[a] In gaseous form, manufacturer's data

[b] AFEAS-Report, September 1989, Ozone Depletion Potential/Global Warming Potential: CFC 11 = 1

[c] Manufacturer's data or Commission report

[d] Estimates

* New model calculations: 0.06 to 0.18

Fig. 91. Open-cell
evacuated rigid foam
sheets welded to gas-tight
facings

specially designed for liquid carbon dioxide. The use of compressed
gases as blowing agents is basically a resuscitation of the "frothing"
process carried out in the 60s with the CFC R-12 (CCl_2F_2, bp:
$-29.8\,°C$). The reaction mix issuing from the mixing head resembles
whipped cream (Fig. 93).

CO_2 froth metering is straightforward, because flexible, semi-rigid
and rigid PU foams can be manufactured both continuously and

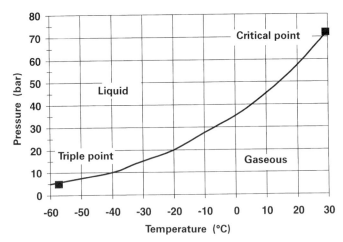

Fig. 92. States of CO_2 : CO_2 can be handled in liquid form under moderate pressure
at room temperature owing to its relatively high critical temperature

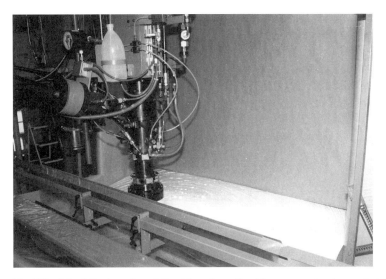

Fig. 93. CO_2 "froth": the relaxed liquid carbon dioxide evaporates immediately; the reaction mix leaving the mixing head resembles "whipped PU cream"

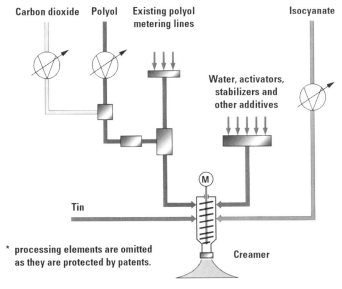

Fig. 94. Metering diagram for producing CO_{2liq}-blown flexible foam. A CO_2/polyol premix is supplied to the existing polyol metering line.

discontinuously. Fig. 94 is a diagram showing the addition of liquid CO_2. Knowledge of the solubility of CO_2 in polyols and isocyanates is required for metering it exactly. Fig. 95 shows proportions of two commercial polyethers and TDI for PU flexible slabstock foam. The quantities of CO_{2liq} used are 2 to 5%, based on a polyol mixture containing 3 to 4.6 parts of water. Low densities which could formerly only be achieved by using R-11 or methylene chloride (CH_2Cl_2) are now feasible, for example a density of $15\,kg/m^3$ [41]:

$$[polyol + 4.6\ parts\ water] +\ \ 5.5\ parts\ CO_{2fl}\ or$$
$$+ 17.0\ parts\ CH_2Cl_2\ or$$
$$+ 21.0\ parts\ R\text{-}11!$$

CO_2 is contained in the air in a proportion of $350\,ml/m^3 =$ ppm $= 691\,mg/m^3$, has a MAC value ($5000\,ml/m^3 =$ ppm $= 9000\,mg/m^3$) which is five times higher than that of R-11 (see above) and is over 50 % heavier than air. It can accumulate close to the ground and has a suffocating effect: so staff must not practice floor exercises! The "dog's cave" in Naples provides a cautionary tale [42]. A much more costly method of blowing makes use of the foaming capacity of the PU mix under vacuum. The amount of foam obtained from a given quantity of raw materials is increased under reduced air pressure. This

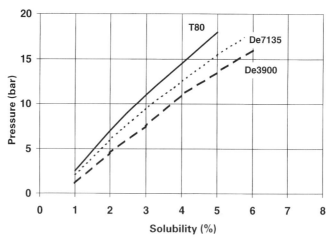

Fig. 95. Solubility of CO_2 at room temperature in PU raw materials for flexible slabstock foams (T-80 = TDI, DE = polyether)

technique was learned from flexible foam production, for example in Mexico City, at 2240 m above sea level in comparison with Germany, which is at a lower altitude. The foaming process ("VPF" – variable pressure foaming) takes place under reduced pressure in vacuum-tight equipment, up to the transverse cutter. This equipment is connected by an air lock to the conveyor belt under normal pressure for finished blocks [43].

Flame retardants are added to reduce the flammability of PU. Inorganic (for example hydroxy aluminium oxides, ammonium polyphosphates) as well as organic chlorine/bromine and/or phosphorus, and occasionally nitrogen-containing compounds, can be used (Fig. 96) [44].

Fillers (for example carbon black, chalk, silicates, heavy spar) can also be used in polyurethanes for "stretching", i.e. to reduce the cost. To

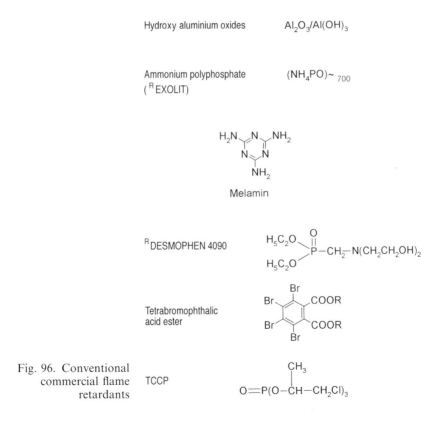

Fig. 96. Conventional commercial flame retardants

Type	Aim	Example
Antioxidants	Delay thermo-oxidation	Sterically hindered phenols
Light stabilizers	Improve stability to light	Benzoxazoles
Hydrolysis stabilizers	Delay hydrolytic decomposition	Polycarbodiimide

Fig. 97. Main industrial anti-ageing agents, mode of operation and chemical basis

improve the physical properties, glass fibres are very important reinforcing materials in structural foams (R-RIM technology).

Anti-Aging-Agents against photooxidation and hydrolysis are also essential for polyurethanes in many applications (Fig. 97).

Colourants for dying the composition are available as doughs or pastes. Preparations made of inorganic or organic dyes or pigments are used in polyols. Dyes which can be incorporated into the PU polymer via NCO-reactive groups are also available [45].

Antistatics, frequently organic ammonium compounds, reduce the electrostatic charge.

Biocides protect polyurethanes from attack by microorganisms (bacteria, fungi).

Release agents for quick and easy removal of PU mouldings from the mould are sometimes contained in the polyol formulation as so-called "internal" release agents. However, "external" surface treatment of the mould is usually essential. Empirical knowledge is required to find the correct release agent; it can have a lasting effect on the cycle time. Waxes, silicones, metal soaps and combinations of these ingredients are commonly used.

6.3.5 Delivery and handling of raw materials; Safety at work

Polyols and di-/polyisocyanates are supplied in drums (200 l), returnable containers (1000 l) or in compartmentalized road tankers (up to 22 t). Construction of a tank farm should be considered if quantities exceed 150 t per year. Plant for the storage and processing of PU raw materials requires government approval, depending on quantity. There are regulations governing transportation, loading and unloading procedures and storage of raw materials, to avoid or at least minimize intrusions into the environment. The safety of employees dealing with raw materials is also controlled by laws and rules (an estimated 500 000 personnel worldwide [46]). Raw materials suppliers provide detailed information, including information on the neutralization of raw material waste, along with an updated safety data sheet on each product [47]. Although pure *polyether and polyester polyols*, unlike natural castor oil (see Section 6.5) which can cause allergies, are usually harmless to health, the normal safety precautions taken when handling chemicals should be observed. This is obvious in the case of polyol formulations which also contain additives. The additives are not always harmless! The reactivity of the NCO group in *di- and polyisocyanates* also affects biological substrates: although it has relatively low oral toxicity, eyes, skin and respiratory tracts must be protected from chemical reactions with the products themselves and from the acute irritation caused by their vapours (for example of TDI) [48]. Prolonged or frequent inhalation of higher concentrations of isocyanate can cause coughing, shortness of breath and sensitization of the respiratory tract. So a maximum workplace concentration of $0.02 \, ml/m^3$ (= ppm) was established to protect employees in many industrial countries just prior to industrial production and marketing of TDI. Large air vents had to be installed in workplaces and foaming plant. Goggles with protective sides, waterproof gloves, impermeable overalls and solid shoes are prerequisites for the responsible handling of isocyanates. Over 30 years' industrial experience with isocyanates involving over a million persons has not indicated prolonged or permanent damage to health by di- and polyisocyanates, apart from occasional sensitization of the respiratory tract.

The carcinogenic potential of isocyanates has been comprehensively tested on animals. Long-term inhalation tests to establish the dangers of TDI had negative results. A standard long-term inhalation test using MDI aerosols led to a few benign lung tumours among rats in the group with maximum exposure of $6 \, mg/m^3$. No tumours were observed with two lower concentrations (0.2 and $1.0 \, mg/m^3$). The official German list of atmospheric limits stipulates $0.05 \, mg/m^3$ for MDI in the form of respirable aerosols [49]. Careful epidemiological investigation of the risk of cancer in persons exposed to isocyanates gave no cause for concern. The results of numerous animal experiments did not show teratogenic or mutagenic effects.

Details about safe handling of the various additives cannot be provided here. The suppliers' technical literature should be consulted (see above).

6.3.6 Trademarks and manufacturers (selection)

Polyurethanes are marketed and used under an enormous number of trademarks. On the other hand, the number of trademarks covering PU raw materials is extensive but limited. Table 27 shows a selection.

6.4   Macromolecular structure of polyurethanes

The title of Chapter 6 indicates that macroscopic examination is no longer adequate for understanding polyurethanes. We now know roughly what to think and do in order to produce a wide variety of polyurethanes. We can pick them up and describe them formally using traditional chemistry. It was pointed out at the beginning of Section 6.1 that *urethane formation* plays the leading role in polyurethanes whereas the *urethane group* only has a small part, for example 4 to 6 % in flexible foams, "real polyurethanes" which come quite close to a

> "poly-{urea-biuret-glycerol-
> [tri-(oxypropylene/oxyethylene block) ether-]}urethane"

An architect can inspect a building macroscopically, but a chemist, as a PU architect, cannot do so: he needs physicists to reveal the inner cohesion of the PU macromolecule using X-rays and electron microscopes.

Even *Otto Bayer's* team of PU inventors in the 1940s used Laue patterns on unstretched and stretched Igamid U, a linear, pure poly-urethane composed of HDI and butanediol-1,4. It has a high degree of crystallization similar to that of polyamides and is still used as a model substance. Recent results of X-ray structural investigations on mono-crystals from a model diurethane (MDI/butanediol) and transmission electron microscope (TME) photographs of a linear TPU (MDI/C_4

Table 27. PU raw materials: selection of a few commercial products and manufacturers

Product	Trademark	Manufacturer
Polyisocyantes	Caradate	Shell
	Desmodur	Bayer
	Isonate	Dow Chem.
	Lupranat	BASF
	Mondur	Bayer (USA)
	Papi	Dow Chem.
	Rubinate	ICI (USA)
	Sumidur	SBU (Japan)
	Suprasec	ICI
	Systanat	BASF
	Takenate	Tekeda
	Tedimon	Montedipe
	Voranate	Dow Chem.
Polyols/systems	Arcol	Arco
	Bay . . .	Bayer
	Caradol	Shell
	Dalto . . .	ICI
	Desmophen	Bayer
	Elasto . . .	Elastogran/BASF
	Glendion	Montedipe
	Lupranol	BASF
	Multranol	Bayer (USA)
	Spec . . .	Dow Chem.
	Sumiphen	SBU (Japan)
	Systol	BASF
	Vora . . . (nol)	Dow Chem.
Thermoplastic polyurethanes (TPU)	Desmopan	Bayer
	Estane	Goodrich
	Pellethane	Dow Chem.
	Texin	Bayer (USA)

ether) lead to an understanding of structure which can also be applied to the majority of amorphous polyurethanes. A distinction is accordingly made between the

- *Primary structure:* it is characterized by the chemical composition of the covalent polymer chain consisting of polyol and di-/polyisocyanate and optionally crosslinking agent/chain extender. The urethane groups and the diisocyanate radical, possibly including the crosslinking agent/chain extender, form the *hard segment* and the (long-chain) polyol radical the *soft segment* (Fig. 98).

- *Secondary structure:* it is characterized by the polyamide-like *interchenar interactions,* for example through hydrogen bridges between the urethane groups of the rigid segments (Fig. 99). π-electron overlaps between parallel aromatic rings of symmetrical diisocyanates are also possible.

- *Tertiary structure:* it is characterized by a certain "order in the chaos" of the polymeric chain coil – as in biological and sociological structures [50] – , in that the rigid and flexible segments are found in *domains.* Fig. 100 shows a TEM photograph on the left and a diagram of a segmented TPU on the right.

Hard and soft segment, hydrogen bridges and domains – as in the foregoing figures – are the "internal fittings" of the macromolecular PU

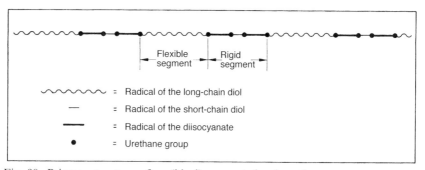

Fig. 98. Primary structure of an (ideal) segmented polyurethane

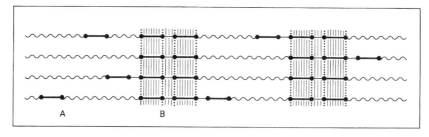

Fig. 99. Interchenar short-range order interaction between rigid segments

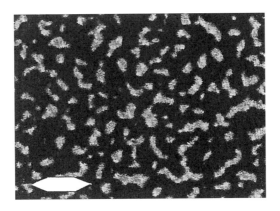

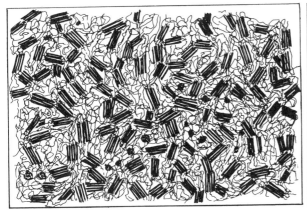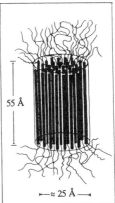

Fig. 100. Tertiary structure: transmission electron microscope (TEM) photograph of a film of segmented PU elastomer (top): the white areas/rods are the cylindrical domains of the rigid segments. The super-lattice with the cylindrical model is shown schematically (right)

building designed by a chemist. For practical reasons, however, paradoxical properties of products can be explained by X-ray structural analysis if, contrary to chemical expectations, the reaction rate decreases rather than increases as the temperature rises. This is the case with pure MDI (melting point 38 C) which dimerizes faster to (undesirable) uretdione in the solid state (see Fig. 62) than in the liquid state above the melting point. The additional expenditure for storage of the pure product is not due to external contamination, as demonstrated by X-ray structural analysis (Fig. 101).

• Diphenylmethane-1,4-diisocyanate (MDI)

OCN—⟨○⟩—CH₂—⟨○⟩—NCO

Melting point: 38°C

• The solid product dimerizes (to uretdione) at room temperature faster than the liquid product stored at 40°C

• The X-ray structure of the crystal shows that the heteromolecular NCO groups are juxtaposed so as to aid uretdione formation

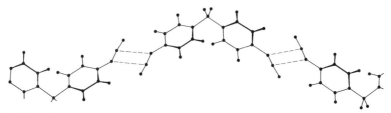

Fig. 101. The molecular cause of the instability in storage of MDI at room temperature was established by X-ray structural analysis

The literature provides accounts of different chemical and morphological structures, including those of standard flexible slabstock and HR foams [51].

6.5   Research and development

At a lecture on polyurethanes, a speaker told the author, as an authority on the subject, that PU chemistry would probably be restricted to NCO and OH groups. However, his perception changed in the course of the discussion. Otherwise, German Patent No. 728961 (1937) (Fig. 102) might have been set aside sooner. These ideas stimulated a tremendous amount of *research,* 60 years ago, into the carriers of the "restrictive" NCO and OH groups. Chemists disrespectfully call these "carriers" a (molecular) "radical", well aware that they determine the reactivity of isocyanates, alcohols, amines, etc. (Fig. 66 and 76). Thus were the *principles* of PU chemistry explored. Research is still the first stage toward innovation but the sequence of ideas has

Fig. 102. The birth
certificate of
polyurethanes:
German Patent No.
728981 of 1937

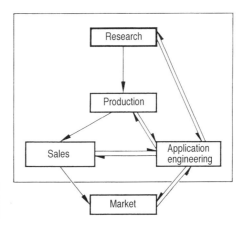

Fig. 103. The flow of ideas is not a
one-way process: R & D are
market orientated

changed. Beyond internal development (see Fig. 110), Fig. 103 shows the striking inter-dependency between the market, marketing (application engineering/sales) and research.

As polyurethanes "matured", the areas of research shifted from basic research through *application-orientated R&D* to *process optimization*. In fact, basic research into polyurethanes, 60 years on, is restricted to a few university investigations on the study of exact molecular structure [52]. These include the most recent chemical rarity, which could be described in PU nomenclature as "ODI", i.e. as *diisocyanate without "radical"*:

$$O=C=N-N=C=O$$

Although the product does not come in a road tanker, its short existence could be demonstrated by vibrational spectroscopy during the photolysis of oxalic acid diazide [53].

The course of the PU *life* ("S") curve (Fig. 9) (still) does not reveal a turning point to asymptotic saturation, but it will be noted that current R&D activity in polyurethanes lies roughly in the restricted region in Fig. 104 after establishment of priorities. The "degree of swelling" of the pear-shaped region is a gauge of the intensity of R&D.

The transition from applied, product-related R&D to process optimization is fluid. An example of this is the changeover from CFCs to other blowing agents: this is associated with a complete revision of *formulation chemistry*, particularly with respect to new synergies between additives [54]. Furthermore, industrial processing must be optimized for handling flammable blowing agents (pentanes, acetone) or liquid carbon dioxide. These trends in development are also reflected in patent literature [55]. The patent literature contains a cornucopia of new "radicals" for *isocyanates*, but they are not likely to compete industrially, economically or ecologically with the above-mentioned standard

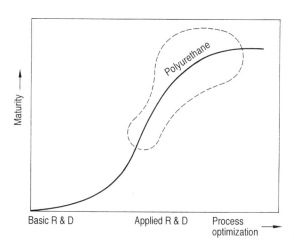

Fig. 104. Diagram of R&D activity as a function of the "maturity" of the product ("S" curve): the focuses for polyurethanes lie in the regions of applied R&D and process optimization

types in the foreseeable future. On the other hand, intensive research and development is aimed at optimizing the processing of isocyanates to reduce environmental pollution and to increase output; it includes processes for phosgene-free synthesis.

The situation has changed with the opening up of new, i.e. *"natural OH sources"*, or, in other words, *renewable raw materials*. Concerted chemical action in the biosphere creates 200 000 000 000 t/a of newly formed biomass of which only 2 to 3% are utilized [56]. These renewable quantities are compared with other resources, including extractions (Table 28).

The "billion ton" ($=10^9$) scale has been selected to illustrate the relationship between supply and demand. Mathematicians can easily calculate the individual ranges of supplies. Beyond the products already used in PU chemistry (see above), the extractions of minute quantities

Table 28. Global supply and extraction of raw materials (in 10^9 = billion tons)

Raw materials	Supply and extraction	Literature
Coal	6940	[7]
currently workable	780	[7]
annual extraction	3	[61]
Lignite	6500	[62]
annual extraction	1.189	
Rock salt (excluding oceans)	3700	[6]
annual extraction	0.2	[6]
Petroleum	626.9	[63]
obtainable by conventional techniques	136.9	[8]
annual extraction	3.261	[8]
for plastics	0.1315	[see p. 21]
for PU	0.0066	[see p. 21]
Biomass (annual production)	200	[56]
annual extraction about 2 to 3%	4 to 6	[56]
oil seeds (1990)	0.218	[64]
soybeans (1989)	0.107	[64]
for soya oil (1988)	0.0154	[65]
castor-oil seeds (1985)	0.0013	[66]
for castor oil	0.000535	[66]
sugar cane/beet	1.3945	[67]
for sugar	0.1103	[67]

of raw materials from plants should encourage researchers and developers to use botanical resources more intensively. Aliphatic substances are preferred; apart from lignin, resources for the synthesis of aromatic substances are less generous. However, we must not expect the natural biomass to replace natural petroleum with regard to plastics [57]. Natural material lobbyists (for example the United Soybean Board/ USA) are also aware of this. They are promoting research into soya protein as a PU polyol component for foams, but with knowledge that the price is (currently) higher than that of traditional polymers [58]. The suitability of sugar beet pulp, lignin, vegetable oils, fats and their chemical derivatives for polyurethanes is also being investigated [58]. Among seed oils, those of the soybean (*glycine hispida*) and the castor-oil plant (*ricinus communis*) are of particular interest (see Table 28). The former is used in epoxide form (ESO) as a plasticizer in addition to its use in food and soap. Soya oil polyols can be obtained by partially opening the oxirane ring with alcohols (Fig. 105) [59].

Castor oil which is known as a laxative (see Section 6.3.5) is not only a raw material for nylon 11, but has always been a (co)polyol used as triol (Fig. 106).

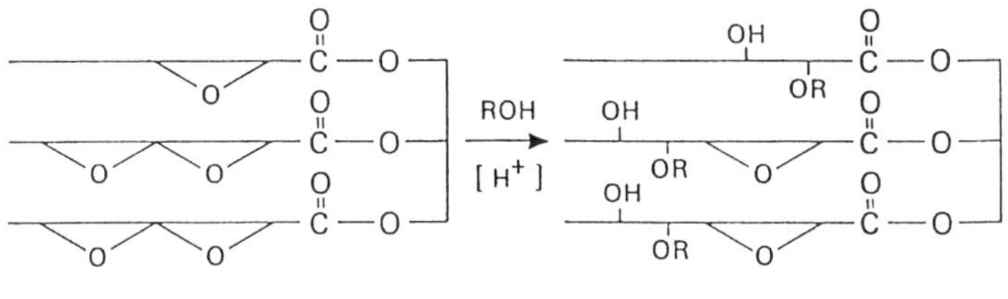

Fig. 105. Obtaining of polyols by partial ring opening of soya oil epoxide with alcohol

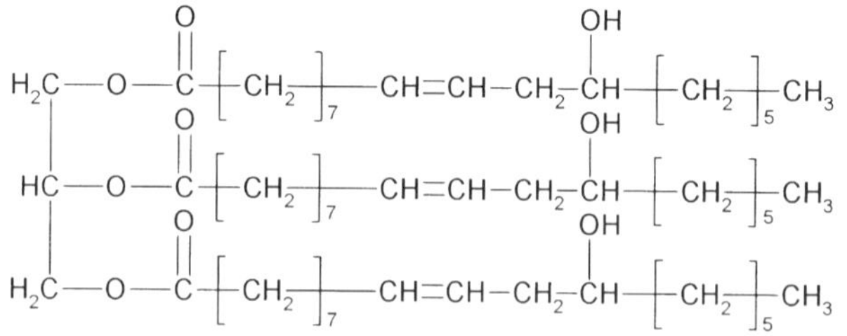

Fig. 106. Idealized structure of castor oil; about 10 % of ricinoleic acid in the natural product (in the figure) are replaced by OH-free fatty acids [60]

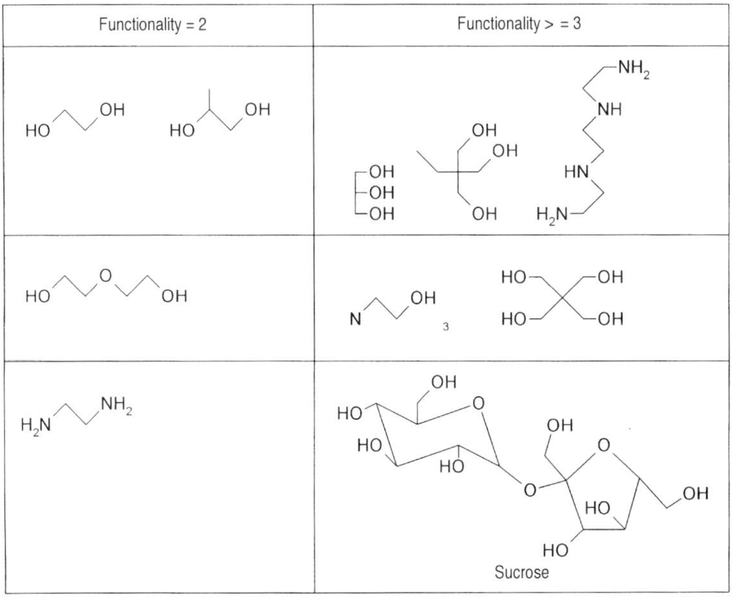

Fig. 107. Products suitable for the ester interchange/amidation of castor oil [60]

A slight drawback is that even the theoretical product (Fig. 106) has quite a low OH value $= 180$. This is even more unfortunate in the natural product as the functionality drops to 2.7 owing to the presence of OH-free fatty acids in the molecule.

Modified castor oils with OH values suitable for the production of rigid foams can be obtained in various viscosities by ester interchange with polyhydric alcohols or amidation with diamines at room temperature (Fig. 107, Table 29) [60].

Table 29. Castor-oil polyols modified by ester interchange [60, p. 322]

Type	OH value (mPa · s/25 C)	Viscosity
1	429	1200
2	410	4500
3	250	450
4	390	250

6.6   Is the chemistry right?
Analysis of polyurethanes

Anyone who chooses polyurethanes on the basis of the information available up until now has been given good advice. The cautious person, who should read Chapters 7 and 8, has been given even better advice. The curious person should pause here to look at the mechanisms behind the chemical procedures.

Why should we analyse polyurethanes?

Firstly, researchers and developers want to know whether the substances which they have concocted are appropriate for a given application.

Secondly, participants in the PU market wish to know what their competitors are cooking up, though contractual "non-analysis agreements" sometimes have to be respected.

Thirdly, the concentrations of PU raw materials which can be detected by trace analysis are of interest, with respect to health and the environment, to industrial doctors and toxicologists, and ultimately to consumers.

Although the composition of polyurethanes can be investigated directly, preliminary chemical decomposition is normally required. Details, for example about isomer compositions of isocyanates, initiators and oxy-alkylene sequences in polyethers or starting materials for polyesters are only available from complex analysis procedures. This applies in particular to the detection of crosslinking agents/chain extenders and the extensive range of additives which often first have to be concentrated by extraction owing to their low concentrations. Modern analysis can also be used to distinguish, for example, between glycerol as a crosslinking agent and/or as an starting alcohol for a polyether and/or as a triol component of a polyester. The analysis of a ready-to-process polyol formulation, for example, is a case in point.

Heating (150 to 240 °C) and pyrolysis at 600 to 700 °C are methods of thermal decomposition, and the fragments can be investigated in solution or in the gaseous phase by means of reference samples. Polyurethanes can be chemically decomposed by hydrolysis. The make-up of polyethers from starting alcohols and epoxides can be obtained by cleavage with HBr (Zeisel cleavage). Individual chemicals and characteristic decomposition products are actually identified almost exclusively by physical or physico-chemical methods. Infrared (IR) and nuclear magnetic resonance (NMR) spectroscopy are the main methods employed. Chromatographic (HPLC, GC, GPC) and spectroscopic couplings (GC/MS) are also normal.

The quantities investigated are minimal and range from a few hundred mg down to a μg, depending on the extent of analysis [69]. Extreme care is required during *trace analysis* if it is considered, for example, that the MAC value of TDI is fixed at 0.01 ppm $(ml/m^3) = 0.07\,mg/m^3$ of air [70]. Indicator band devices which continuously monitor the ambient air are suitable, for example, in TDI-processing facilities, whereas a sampling strategy (place, time and duration) is an excellent way of obtaining correct test results in critical cases. Low readings could endanger affected persons and could be fatal. On the other hand, mistakenly high values would lead to industrial precautions against dangers which do not actually exist.

Compulsory safety equipment for employees involved in isocyanate production includes an indicator plaque [6] which is worn on overalls and displays critical phosgene concentrations ($COCl_2$: MAC = 0.1 ppm = $0.4\,mg/m^3$ air).

Environmentally hazardous products such as CFCs can also be specifically detected. Blowing gas can be extracted from the cells of old rigid foams using gas-tight syringes and can be investigated directly by gas chromatography to detect R-11 ($CFCl_3$; MAC = 1000 ppm = $5600\,mg/m^3$ air).

7   The quality must be right

Are we referring to subjective or objective quality? Neither! Commercial success depends only on the *relative quality*. It is determined by comparing the qualities of products and services offered by competitors. The *contractual quality* between sellers and buyers is also decisive. Finally, the end user will make a subjective judgement about the quality of his purchase by comparing price and performance. He will note that top quality cannot be obtained at the lowest price and he should be pleased with *consistently high quality* at reasonable prices. This pleasure (and therefore the *quality of life?*) can be marred if the end user has to pay to modify or dispose of his purchase in an ecologically acceptable manner. In other words, quality includes an ecological parameter on top of price and performance.

In addition to value for money, politics and society will play a greater role in determining whether and how a product fits in. Ethical aspects will also determine its acceptability [71]. These factors lead to the tetrahedral model of "new quality" (Fig. 108) [72].

As the polyurethane market (Fig. 7) is split between PU raw material manufacturers, PU plant builders and PU manufacturers, total quality management is required. While raw material manufacturers *guarantee* specified polyisocyanate and polyol component data (Table 30) and *machine builders* guarantee the reliability of their PU plant, PU

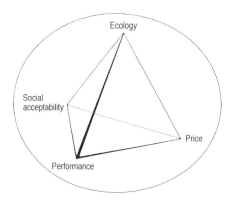

Fig. 108. The tetrahedral model of "new quality": if "performance" and "price" are replaced by *technology and economy* and if Fig. 12 is supplemented by the term *technology*, "new quality (of life?)" becomes synonymous with "sustainable development" [72]

Table 30. Chemical and physical characteristics affecting the processing of PU raw materials

Property Raw material	Purity or NCO content %	OHV mg KOH/g	AV mg KOH/g	Visc. MPa·s	H₂O %	Density g/cm³	Miscellaneous, for example			
							pH	Hydrol. Cl. %	Sediment %	PHI*) content ppm
Di-/(poly-)isocyanates	X			X		X		X	[X]	[X]
Polyether polyols	X	[X]	X	X	X	X	X			
Polyester polyols		X	X	X						

*) phenylisocyanate

manufacturers have to comply with processing parameters. As suppliers to the automotive industry, for example, they assure their customers that parts are of the required quality. For example [73], the PU properties mentioned in raw material manufacturer's technical documents are often given as a guide and not as binding minimum values of the finished part. The same applies to the numerical details in this book. The PU manufacturer's mould design and processing conditions can have a decisive influence on the quality.

The construction of a PU total quality management system involves internal and external audits. Quality assurance is based on the standards

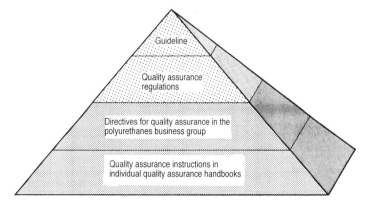

Fig. 109. Example of the hierarchical organization of a QM system

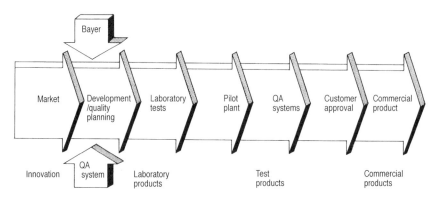

Fig. 110. QM criteria are taken into consideration in the early stages of development of a new PU raw material

EN 29000 to EN 29004. Immaterial quality management procedures (relating to services) are included in addition to material procedures (relating to products). Without going into detail, the main features of a corporate QM system will be demonstrated by reference to a major PU raw material manufacturer [74]. The system includes elements from the publication "Quality Assurance in the Chemical Industry" published by the Chemical Industries Association (VCI) (Fig. 109).

The *QM Guidelines* which are issued by the Board and are binding for the firm are converted into *QM regulations* specific to the Polyurethanes Group and are set down as QM *directives* geared to Research, Application engineering, Production and Sales. Customer requirements can also be included at this level. The *QM instructions* which are binding for each employee are derived from these guidelines and are set out in specific *QM handbooks*. Fig. 110 shows early incorporation of the QM system into the development of a PU raw material. Encouragement for product development can also emerge, for example, from batch control.

8   Waste and recycling

People who thought that unlimited prosperity lay in natural resources had a shock in 1973/74: the quadrupling of the price of petroleum from 2.50 to over 10 US$ per barrel ("first oil crisis") as a result of the fourth Israeli/Arab war. Although oil consumption rose by over 12 % in the following six years and the oil price has more than trebled at about 34 US$ per barrel ("second oil crisis"), before consumption and price fell back to the – higher – level of the first "oil shock", this aroused fear and anxiety and then awareness. A wise man stated: *"This psychological shock has been a blessing for us all. We must now leave our anger and fear behind us"* [75]. Normal consumers and their representatives eventually had to learn how to save energy – and this has become a pillar of chemical engineering though "only" for economic reasons. The concept of the "calorie hunter" is a sign of the times [76]. The modern variant is the energy loop which includes end products, energy recovery and ash (not shown in Fig. 111):

The Closed Substance Cycle and Waste Management Act which came into force in Germany on 7th October 1996, and a range of new ordinances (Table 31) [77] are directly related to this.

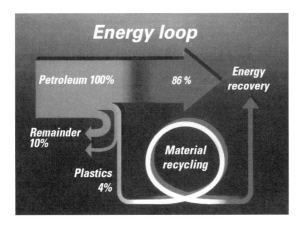

Fig. 111. Plastics represent the "interest" on the 4 % of petroleum "loaned": after use and recycling, the borrowed resource-capital is returned as energy-profit

Table 31. Extract from the bundle of ordinances for enforcement of the Closed Substance Cycle Law [78]

Ordinance on the Classification of Waste
Ordinance on the Classification and Recovery of Waste
Ordinance on the Furnishing of Proof
Ordinance on the Transport of Waste
Ordinance on Specialized Waste Management Companies
Ordinance of the Introduction of the European Waste Management Catalogue
Ordinance on Waste Management Concepts and Waste Life-cycle Analysis
Directive on Waste Management Partnerships

The law is geared to the production of *low-waste, durable, reusable, easily repaired or at least easily recycled* products. Customers must focus on durable consumer goods. Legislators realize that waste cannot be completely avoided. Finally, damage to the polymer network by repeated processing prevents unlimited recycling. In the final analysis, the second basic principle of thermodynamics applies here. There are still some waste residues which have to be treated in environmentally acceptable ways. These include high-quality energy recovery governed by the new closed substance cycle law, known as combustion with recovery of energy. Small landfill site capacities will therefore last longer. Our purses tell us that this is "worthwhile" (original Federal Ministry for the Environment soundtrack): the cost of waste disposal has more than doubled in Germany since 1990. If less waste is produced (1990 to 1993, about -37 million tons $\sim -10\%$), the proportion of fixed costs per ton of refuse will increase. The number of landfill sites in Germany has dropped to 562 and further closures are anticipated. Since the Closed Substance Cycle and Waste Management Act and the Technical Directive on domestic Waste, earth-like materials only may be dumped as from the transitional period between 2001 and 2006. Organic material must be pretreated, i.e. incinerated. The current 52 household refuse incineration plants will be inadequate, so additional plant will have to be constructed [78]. The current excess capacity in incinerators or landfill sites are a temporary phenomenon when viewed in the longer term, but a lucrative playground for the waste disposal industry for the time being [79].

How do polyurethanes fit into the legal domain?

Used correctly, PU end products do not harm the environment [80] (see

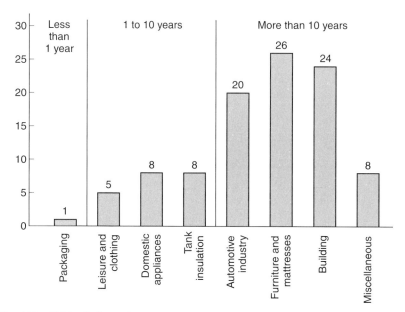

Fig. 112. 78 % of all PU is used in durable products

Section 4.3). PU generates little waste, i.e. is more durable than PU-containing products (Fig. 112). PU scrap is formed at the end of the product's life cycle and during its production, for example as a blend, sprue waste, reject mouldings, etc. Contrary to popular opinion, this PU scrap can be recycled. Fig. 113 shows the most important methods of recycling PU scrap [81]. It should be remembered that the methods are not all suitable for every type of PU. Finally, recycling should not become an end in itself. Even recycling consumes energy, for example surface energy for shredding the waste usually occurring in coarse lumps in addition to transport and sorting energy. When viewed as a whole (see also Fig. 108 and Fig. 12),

> industrial (feasibility, applications),
> ecological (acceptability),
> sociological (acceptance) and
> economic (recycled material market)

aspects have to be suited to each other. These are key parameters for sustainable development (see also Fig. 12) through responsible care by

Recycling of polyurethanes		
Material recycling	Raw material recycling	Energy recovery
Flocking	Glycolysis	Household refuse incineration
Adhesive pressing	Hydrolysis	Rotary kiln
Particle bonding	Pyrolysis	Smouldering process
Powder incorporation	Hydrogenation	Metallurgical recycling
Injection moulding	Glass production	Industrial power station
Flow pressing	Reduction	

Fig. 113. Polyurethanes can be recycled – just like other materials

all active and passive participants, i.e. manufacturers and consumers. It can be inferred from this that the recovery quota for used plastics, and also for PU, cannot be set as high as we want. Fig. 114 shows the theoretical relationships – though they have been overdrawn. Curve b is the "problem curve". If the last PU mattress were collected from who knows where as the result of an unrealistic 100 % collection order, the energy expended in transport could increase exponentially and make a mockery of responsible care. With a recovery quota of about 40 % there is a real saving in energy. However, data currently available is not sufficient for a realistic representation. We do not know the source of the recycled raw material or, in other words, where does the PU waste come from? The quantities shown in Fig. 115 are possible expectations from the specified sources of waste. Interpretation does not give rise to euphoria. Without going into detail, it can easily be seen that relatively small quantities can be highly dispersed geographically so the choice of recycling concept is not so simple. These relationships illustrate the problems encountered, in the case of the disposal of (PU) refrigeration appliances with an estimated service life of 19 years, when recycling about 2.6 kg of PU rigid foam per old refrigerator in compliance with the above-mentioned key parameters. About 3000 t/a of degassed (R-11) PU rigid foam powder are being produced by the disposal of old refrigerators in the 1990s. 15 000 t/a are anticipated at the turn of the century as more, larger appliances will be produced. The

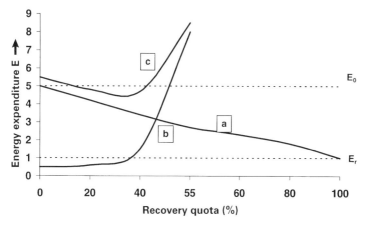

Fig. 114. Simplified diagram of the recovery quota for plastic waste. It is about 40 %.
E_0 Energy for production from virgin material
E_r Energy for production from recycled material
(a) Energy for production according to content of recycled material
(b) Energy for collection, sorting, transport
(c) Total energy according to content of recycled material

Flexible foams	
— From furniture and mattresses:	about 30,000 t / a in bulky refuse about 15,000 t / a from hospitals
— From used car seating	about 35,000 t / a from dismantling factories (about 175 t / a per factory?)
Rigid foams — From domestic appliances	about 9,000 t / a from refrigerator disposal firms
— From factory buildings	? from industrial areas
— From residential buildings	? sorted from rubble
RIM polyurethanes — From vehicle repair	about 1,500 t / a from the automobile industry

Fig. 115. Estimate of the maximum possible quantities of PU waste from selected sources for the period between 1995 and 2000, over the entire country (without exports!)

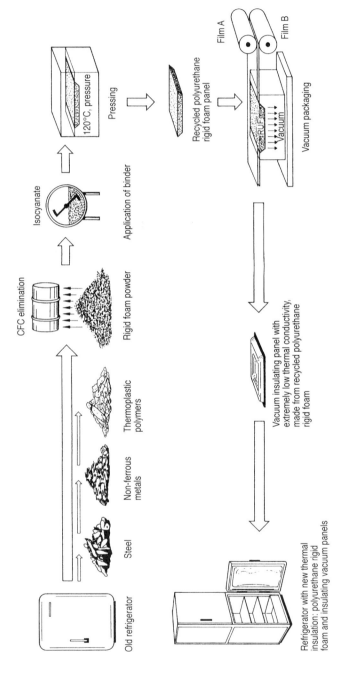

Fig. 116. Recycling of PU rigid foam from an old refrigerator to a new one

majority is still incinerated or dumped nowadays. Old rigid foam powder can also be returned to its original purpose by bonding it with PMDI and compressing it to form open-cell rigid foam panels. These panels are welded in foil under vacuum and can be installed in new refrigerators as insulating vacuum panels (cf. Section 6.3.4) with a little fresh PU rigid foam (Fig. 116). At a vacuum of 0.7 mbar, the thermal conductivity of the insulating panels is 8 to $10\,\text{mW}\cdot\text{m}^{-1}\cdot\text{K}^{-1}$ in comparison with 20 to $21\,\text{mW}\cdot\text{m}^{-1}\cdot\text{K}^{-1}$ for rigid foam blown with cyclopentane. This is an example of almost "1:1 recycling" of polyurethane [82]. The circumstances described up until now may just be tiny stones in the overall psycho-technical mosaic of "environmental protection" – obviously including PU. This pattern of stones is also called "ecological balance" or "product line analysis"; foaming and casting practitioners speak of "cradle to grave analysis". The modern term is "LCA": life cycle analysis [83]. Fig. 117 shows the basic elements of an ecological balance. Its usefulness in helping with decisions on environmental protection is not disputed. The enormous amount of time and money required to create and evaluate a comprehensive industrial data pool, including psychological imponderables (cf. also Fig. 108 and 12) is in dispute. A PU ecological balance would definitely be nonsensical; individual LCAs based, for example, on PU raw materials via PU mattresses, PU-insulated refrigerators, etc. are worthwhile. The associations (see Section 9.1) have made a start and preliminary results are available (for example material balances for

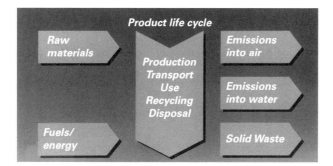

Fig. 117. These elements of the life cycle are universal – possibly with different names. They apply to the contents of all organic life right up to the materials; we call them the "ecological balance" in this case

MDI, TDI, polyethers) [84]. However, it should be remembered that life cycle analysis is not a panacea or guarantee of the correct procedure. The Federal Environment Agency states: "A method of evaluation which can comply with the standards of natural science and industry and is frequently requested by the public does not exist" [85]. The aim of supplementing the overall economic equation (social product) with the standard variation of "natural capital", i.e. the environmental audit, cannot be achieved in the near future either [86].

Recycling of materials can be carried out by PU manufacturers themselves. Examples include flocking created from PU flexible foam production waste: between 5000 and 10 000 t/a is marketed in Germany alone and about 20 000 t/a in Western Europe, for example for gym mats, special cushions, sound absorbing elements, etc.; compare the United States, where 200 000 t/a are marketed! This includes about 4000 to 6000 t/a of pressed sheets created from shredded PU rigid foam scrap. 200 to 300 t/a of granulated RIM polyurethane from automotive

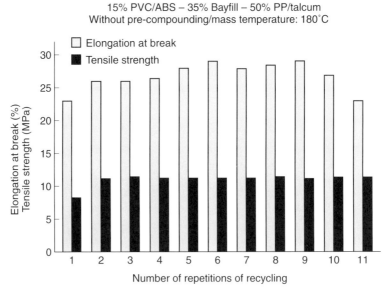

Fig. 118. Surprising properties as a function of the frequency of recycling: elongation at break and tensile strength after repeated thermoplastic processing of dashboard material

applications are used for specific building applications. The granulated material produced from car dashboards containing 35 % of PU
foam and 65 % of ("auxiliary") thermoplastics (PVC/ABS/PP) can be
processed on injection moulding machinery and – what a surprise – its
mechanical properties are not only unchanged but are improved up to
the sixth processing treatment during ten-fold recyclate processing: is a
new morphology being created here (Fig. 118)?

Instead of (flexible foam) flakes, any type of shredded polyurethane can
be processed into *particle boards*. If this is carried out under pressure
and heat, the experts call it *adhesive pressing*. The *powder* (Fig. 120)
obtained by milling (Fig. 119) flexible foam can be incorporated into
fresh polyol as filler and can be processed with isocyanate to form a new
part or slabstock material (Fig. 121). In a first stage the foam blend is
shredded into flocks which, in a second stage, are processed by high
shearing forces in roll mills, possibly by repeated grinding into powder
with an average particle size of 200 μm which can be stored in silos. By
means of a stuffing screw – a technique transferred from the metering of
melamine for CMHR foams – the foam powder is fed continuously via
a premixer to the polyol stream (differential stream system) and then
mixed with isocyanate and possibly additives in the mixing head.

Fig. 119. Roll mills shred
PU foam scrap into
powder, possibly in several
passes (flexible foam
regrind powder)

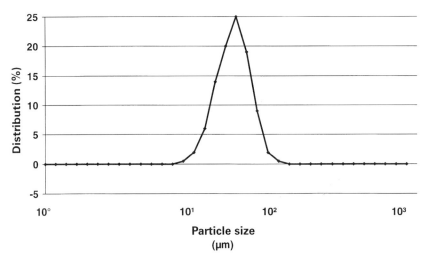

Fig. 120. Particle size distribution (% and m) of PU flexible foam regrind powder by milling

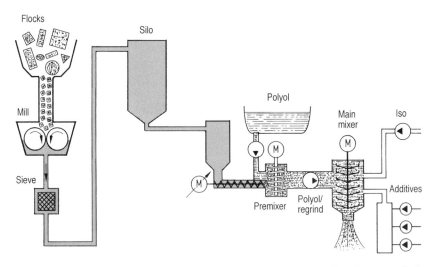

Fig. 121. General view of production and incorporation of flexible foam regrind powder into the reaction mix

In the case of polyurethanes with a lower level of crosslinking (flexible to semi-rigid qualities), their "thermoplastoid" properties are utilized (see Section 4.1). Preheated granulated material can be pressed into mouldings at about 180 °C and 350 bar by *flow pressing* in addition to *injection moulding* (see above). The granular particles merge superficially and adhere without additional binder.

Detailed knowledge of chemistry is required when *recycling raw materials.* In the final analysis, the PU-containing polymer molecule has to be destroyed again, with the consumption of energy. Chemical apparatus is required for this purpose, whether for hydrolysis, glycolysis, aminolysis or pyrolysis, or for hydrogenation or gas production [87]. Different low-molecular weight fragments are formed, depending on the method employed. Industrial capacities are already available for this purpose. Reference will be made to gylcolysate production as an example of the recycling to raw materials (Fig. 122).

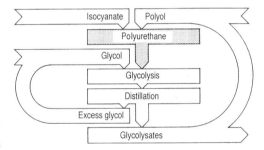

Fig. 122. Diagram of PU
glycolysis

In the past, this was the only raw material recycling carried out on polyurethanes. A quantity of about 800 t was estimated in Western Europe for 1994 for various applications. Polyether polyols, short-chain, OH- or NH_2-terminated urethanes and traces of aromatic amines, which have to be allowed for when reprocessing the glyco-lysate, are primary products of glycolysis. Residual amounts of aromatic amines (MDA, TDA) in the recycled polyol can be chemi-cally neutralized, for example by isocyanates or epoxides. A PU rigid foam manufacturer will mix his foam scrap with purchased polyester scrap from PET bottles, textiles, etc. and glycolyse the mixture with a specially developed catalyst to form a recycled polyol, which is recirculated to the production of PU rigid foam [88].

The cooperation between a PU raw material manufacturer, three processors, a chemical recycling firm and a car manufacturer provides an example of sustainable development through responsible care. *Material and raw material recycling are combined*: the offcuts created

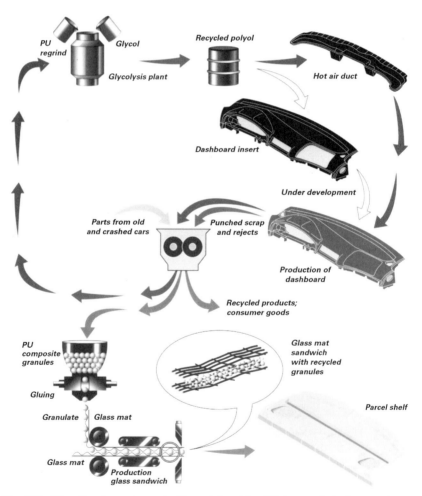

Fig. 123. The all-polyurethane dashboard (see Fig. 17) is recycled after use into a different part in terms of material but to the same part in terms of raw materials

when producing a dashboard (see Fig. 17) made completely of polyurethane are processed as a material to form a parcel shelf and as a raw
material via a recycled polyol to form the hot air duct of a car – an
example of the segregation so important for recycling [89] (Fig. 123).

This is only one of many examples: ISOPA (cf. Section 9.1) publishes
up-to-date lists of firms involved in PU recycling [90].

Depending on regional factors, *energy recovery* (for example combustion) can be carried out in modern plant with safe emission limits.
However, detoxification of the flue gas, which represents progress in air
purification creates filter dust which has to be disposed of as toxic
waste. Nothing is free – disposal is possible but involves costs. There is
a 99 % reduction in volume during the combustion of PU foam – an
incomparable advantage, for example in relation to the immense
quantities of rubble in the form of a mineral building material which
cannot be reduced in volume by combustion. Of 85 million t/a rubble,
31 million t are recycled and 54 million t dumped; of these, 8 million t
contain toxic substances [91]. Only about 0.2 % of annual petroleum
extraction is used for polyurethane (see Chapter 2), but this is still 6.6
million t which, when processed into polyurethane, has a calorific value
close to that of coal (Table 32).

Table 32. Net calorific values of fossil and renewable fuels in comparison with
plastics and waste
(Source: *K. Kroesen*, ICI, in *Raßhofer*: Recycling von Polyurethan-Kunststoffen.
Hüthig, Heidelberg 1994, p. 279)

		Net calorific value MJ/kg
Fuels	Fuel oil	42
	Coal	30
	Lignite briquettes	20
	Wood	15
Plastics	Polyethylene, polypropylene	44
	Polyurethanes	28
	Polyvinylchloride	18
Waste	Used tyres	30
	Used paper	16
	Shredder waste	13

It would obviously be a sin in terms of energy to dump PU waste. It can be incinerated along with domestic or industrial waste, as it is in relatively small quantities. On the other hand, the results of a large-scale incineration test in a stationary coal-driven fluidized bed reactor with 21 t of powdered flexible foam demonstrated that an extra 500 to 700 DM/t of powdered PU waste had to be paid, depending on the throughput – without transport and dismantling costs for the end product [92]. The spare capacity in waste incinerators in the chemical industry can obviously also be used at commercial prices [93]. The CFC R-11 (monofluorotrichloromethane) contained in old PU rigid foams is also broken down by 99.9 %. The chlorine mobilized in this way does not increase the content of polychlorinated dibenzodioxins (PCDD)/furans (PCDF). Rather, state of the art refuse incinerators act as dioxin sinks [94]. Owing to its high nitrogen content, PU does not cause higher NO_x emission than coal. However, the pressure of "public opinion", which often originates from obliging individuals who are not technically minded, sometimes prevails over what is ecologically sensible. Organizations can also mould public opinion, if the horror content is sufficiently high: with hindsight, we can say we are sorry we made a mistake. For car-driving, clean world enthusiasts, the following comparison should provide a constructive aid:

Air pollution by cars ($g \times car^{-1} \times km^{-1}$) and legitimate incineration of the statistical 0.8 kg of refuse per head in incinerators is compared. Industrial possibilities cannot always be exploited fully for reasons of cost, logistical problems (for example collection, separation, identification, sorting). In addition, the amount of PU scrap is quite low owing to the durability of polyurethanes. On the other hand, the receptiveness of the market toward recycled materials and products, which is limited

Table 33. Comparison of car and incinerator emissions
(Source: *K. Kroesen*, ICI, in *Raßhofer*: Recycling von Kunststoffen. Hüthig, Heidelbert 1994, p. 280)

	from cars g/car/km	Incineration of 800 g of refuse g
No_x	0.95	0.880
HC (hydrocarbon)	1.17	0.044
CO (carbon monoxide)	6.12	0.220

in any case, cannot be improved in the short term. Against this background, the dumping of PU scrap on landfill sites is still unavoidable but is only a fraction of the 1.3 million t of plastic scrap forcibly dumped in 1992 [95]!

The above-mentioned protagonists should be aware that energy is wasted in this way and "dead" capital produced. Despite the undisputable technical progress in waste treatment and the management of energy, however, psychological acceptance of public development work is required as enthusiasm for collection alone will not solve the problem. 80 million Germans still cause higher environmental pollution than 900 million Indians [96].

9 Polyurethane know-how

Curious readers can obtain further information on any aspect of PU from the following sources. Section 9.1 refers to manufacturers, associations and exhibitions, i.e. to places to go, to write to, to phone and to fax for those in a hurry. Those who have more time can consult the PU literature mentioned in Section 9.2.

9.1 Who knows what?

Annual directory of PU raw material and PU plant manufacturers with product ranges:

- David Reed (Editor): "Global Polyurethane Industry Directory" 1997 edition, 144 pages, Crain Communication Ltd., 4th Floor New Garden House, 78 Hatton Garden, London EC1N 8JQ, UK; Tel. 0044-(0)171457-1400; Fax 0044-(0)171457-1440

International PU Congresses

- UTECH every two years in The Hague; organizer: Crain Communications. *Abbreviation* for quotations from "UTECH '96 The International Polyurethane Industry Conference, 26-28th March 1996, The Hague, The Netherlands, Book of Papers". Crain Communications Ltd., London (*UTECH '96 in bibliography*)
- UTECH Asia, every two years in different locations; organizer: see above
- PU World Congress, every three years in different locations; organizer: ISOPA and Society of Plastics Industry (SPI)

Associations

- ISOPA European *Isocyanate Producers Association* Avenue E. van Nieuwenhuyse 4, Box 9, B-1160 Brussels, Tel. 0032-2-676-7475, Fax 0032-2-676-7479, members of ISOPA are ARCO, BASF, BAYER, DOW, ENICHEM, ICI and SHELL. ISOPA is a group within the CEFIC (European Chemical Industry Council).
- III *International Isocyanate Institute*, Inc. 201 Main Street, Suite 403,

La Crosse, WI 54601 USA, with a European branch at: Gilbert International Isocyanates Scientific Office, Bridgewater House Floor 9, Whitworth Street, Manchester M1 6LT UK; Tel. 0044-161-236-3774; Fax 0044-161-236-7855

- EUROPUR European Association of *Flexible Polyurethane Foam Blocks* Manufacturers, c/o FIC, Square Marie-Louise, 49, B-1000 Brussels, Tel. 0032-2-238-9869; Fax 0032-2-238-9998

- BING Federation of European *Polyurethane Rigid Foam* Associations, Kriegerstraße 17, D-70191 Stuttgart; Tel. 0049-711-29-1716; Fax 0049-711-29-1716-4902

9.2   What is where available?

Comprehensive overview of polyurethanes with numerous quotations from the literature and patents:

- *G. Oertel* (Editor): Polyurethane Handbook, second edition, Hanser Publishers, Munich, Vienna, New York 1994; 688 pages

Comprehensive overview of polyurethane recycling with numerous quotations from literature and patents:

- *W. Raßhofer* (Editor): Recycling von Polyurethan-Kunststoffen. Hüthig Verlag Heidelberg 1994; 440 pages.

Summary of polyurethanes (chemistry, forms, process engineering, applications, quality assurance and recycling) in:

- *Hj. Saechtling, K. Oberbach* (Editors): Kunststoff-Taschenbuch, 26th edition, 967 pages. Contributions by *K. Uhlig*, "Polyurethan-(PUR-)Verfahrenstechnik" (page 286) and "Polyurethane" (page 491). Carl Hanser Verlag, Munich, Vienna 1995.

Summary of PU foams and elastomers:

- *R. Leppkes:* Polyurethane – Werkstoff mit vielen Gesichtern. Verlag Moderne Industrie, Landsberg 1993, 70 pages.

The only periodical in the world relating specifically to polyurethanes with weekly updating of average market prices of the most important PU raw materials; frequency of publication: bi-monthly (even months); size about 40 pages:

• *David Reed* (Editor): Urethanes Technology. Crain Communication Ltd, London (*abbreviated to UT in the bibliography*)

Company-specific description of polyurethanes

• *G. Woods*: The ICI Polyurethanes Book, 2nd Edition, ICI Polyurethanes Everberg, Belgium jointly published with John Wiley & Sons, Chichester 1990, 364 pages.

Scientific-technical summary of polyurethanes 285 literature references:

• *D. Dieterich, K. Uhlig*: Polyurethanes, In Ullmans Encyclopedia of Industrial Chemistry, vol. A 21. VCH Verlags GmbH, Weinheim 1992, pages 665 to 716.

10   Prospects

Somewhere between euphoria on the one hand and consternation on the other, there is a certain type of man who we will call a real optimist. He avoids mere "growth" as well as the fear of decisions typical of the "zero risk" mentality. The term "development" is better than "growth" for drawing conclusions from the past and present about the future. An upturn and a decline are included in the evaluation of risks. We know that unrestricted growth causes disease and, from prehistoric experience, the "zero risk" is Utopia.

To evaluate the development of polyurethanes in the foreseeable future, we will first bring ourselves up to date by looking at Fig. 8, 9 and 10 in Chapter 2. For decades, polyurethanes have made up about 5 % of total plastics consumption throughout the world, i.e. they have grown proportionally with plastics. The author has no reason to assume that this will change much. As already mentioned elsewhere, the global PU consumption curve fortunately shows no sign of turning. However, the years before the oil crises and at the end of the 80s show how unreliable linearly extrapolated prognoses can be. Of these slumps, which were then considered catastrophic, only lumpy patches appear in the quarter-century curve. Worry about the decline in polyurethanes in the early 90s, right into the negative range in Germany (Fig. 4), soon gave way to joy about the growth in the mid 90s. The challenge for the future is to plan and continue this curve and its first derivative (Fig. 4) with real optimism. The polyurethanes (Fig. 13) themselves and the diversity of applications (Fig. 10) are good reason for this; however, the success of polyurethanes also depends on their wellbeing. Furthermore, different geographic areas will develop differently. For example, two-figure growth rates are expected for PU foams in the USA market between 1994 and 2000 [98]. Europe is confronting three challenges which will eventually be transferred to other industrial regions [97]:

- Maintenance of *competitiveness* by reducing costs owing to the opening of national boundaries. For example, logistics, factory size and integration are more important than labour costs for the few large raw material manufacturers.
- Acceleration of R&D, better known as *creativity and innovation*, with

a view to improving the quality of products and processes and to developing new application areas (cf. Section 6.5, Fig. 87).

- *Environmentally responsible sustainability*, and convincing the world outside the factory wall of it.

With a little imagination, these challenges are reflected in Fig. 12 and Fig. 108. With all the checks required, however, nothing will happen without a minimum of trust between individuals. On this basis, we can be optimistic about the future, and about the polyurethane age.

Polyurethanes were once described by a journalist as the "jack of all trades" among plastics [99]. Specialists with a sense of humour had already unmistakably portrayed this popular classification graphically (Fig. 124).

Fig. 124. Polyurethane: the "Jack of all trades"

Bibliography

[1] International Furniture Exhibition, Cologne 13–19.1.1997; report in special edition of, "Kölner Stadt-Anzeiger" No. 12, 15.1.97, p. 4

[2] *A. Wurtz*: C. R. Hebd. Seances Acd. Sci. 27 (1848), 241

[3] RÖMPP Chem. Lex. 6th Edition 1992, p. 4838

[4] "Urethane Technology"

[5] UT April/May 1996, p. 1

[6] "Chemie mit Chlor – Chancen – Risiken – Perspektiven", Bayer AG, April 1995, p. 9

[7] *Römpp, H.:* Römpp Chemielexikon, 9th Edition, vol. 5, Thieme, Stuttgart 1992, p. 4292

[8] ESSO, OELDORADO '96 ref. In Nachr. Chem. Tech. Lab. 44 (1996) No. 10, p. 1009

[9] Kölner Stadt-Anzeiger No. 231, 3rd/4th October 1996

[10] *Frank Waskow,* on the board of the Cologne Environmental Institute "Katalyse e.V" notices, when buying carpets, that "polyurethane foam has not been used for the backing". Kölner Stadtanzeiger of 4.1.1996, p. 14

[11] KstA No. 231, 3/4.10.1996, p. 60

[12] UTECH '96, PAPER 26

[13] Press release 18/96 UBA

[14] UT, Feb./March 1996, p. 19

[15] UT, Feb./March 1996, p. 46

[16] "Hennecke Connect", Technical Data Sheet TD 507, Maschinenfabrik Hennecke GmbH, St. Augustin (Birlinghoven)

[17] UTECH '96, PAPER 20)

[18] UTECH '96, PAPER 16

[19] UTECH '96, PAPER 49)

[20] DE-OS 19 510 652 (1994) Bayer AG

[21] UT Feb./March 1997, p. 44

[22] *D. Dieterich* in *H. X. Xiao, K. C. Frisch* (Ed.): Advances in Urethane Ionomers, Technomic Publ. Co. Inc, Lancaster, Basel 1995, p. 1

[23] CHEM. RUNDSCH. 49 (1.3.1996) No. 9, BASF

[24] *K. Uhlig:* PU World Congress Aachen 1987, Kunststoffe 78 (1988), p. 73 and UTECH '96, PAPER 43

[25] UTECH '96, PAPER 46

[26] *T. L. Fishback, C. J. Reichel* (BASF Corp.), in [60], p. 282

[27] Poly Bd document 1504-FED/12-91/40 by Elf Atochem ATO Deutschland GmbH, Düsseldorf, and UTECH '96, PAPER 42

[28] Kraton Liquid L 2203, from the document "Kraton Polymers" by Shell Chemicals Europe, London, 1996

[29] Epol by Atochem Indemitsu in UTECH '96, PAPER 45

[30] UTECH '96, PAPER 35

[31] *W. Wacker* (Bosch), *A. Cappella, W. Hoffmann* (ICI), *F. Özkadi, S. Odabas, S. Algan, M. Y. Tanes* (ARCELIK A. S.), *F. Xinghe* (Jingsu Prov. Res. Inst.

Chem. Ind.), *X. Xiaofeng* (Kagoshima Univ.), *J. Murphy* (Elf Atochem NA) in [60], p. 18, 62, 411

[32] UT Dec. 95/Jan. 96, p. 28

[33] UT Aug./Sep. 1996, p. 42–50

[34] *H. P. Doerge, M. H. Venegas* (BAYER Corp.) in [60], p. 24, 57

[35] *I. E. McGee, M. A. Dobransky, K. A. Ingold, F. C. Rossitto, M. E. McGregor* (BAYER Corp.), *M. C. Bogdan, D. J. Williams, P. B. Logsdon, R. C. Parker* (AlliedSignal Inc.), in [60], p. 170, 394, 434, 448

[36] *J. B. Blackwell, G. Buckley, B. Blackwell* (Beamech Grozp Ltd.), *M. Taverna, H. Meloth* (Cannon Group), *T. Griffiths* (CarDio BV), *D. V. Mariani* (Impianti OMS Group) in [60], p. 136, 164, 370, 376

[37] UT June/July 1996, p. 36

[38] BMU (Federal Ministry of the Environment) press release 42/97 of 17.7.1997, p. 3

[39] UTECH '96, PAPER 53, *P. P. Barthelemy, A. Leroy* (Solvay Res. & Tech.) in [60], p. 404

[40] UT April/May 1996, p. 34

[41] UTECH '96, Papers 10, 30 and 31

[42] *Holleman/Wiberg:* Lehrbuch der Anorganischen Chemie. Walter de Gruyter & Co., Berlin 1952, p. 296

[43] UTECH '96, PAPER 32

[44] *R. S. Rose, L. J. Likens, J. L. Martin* (Great Lakes Chem. Corp.), in [60], p. 325

[45] UT Feb./March 1996, p. 30

[46] *F. K. Brochhagen:* Isocyanates. In *O. Hutzinger* (Ed.): The Handbook of Environmental Chemistry, vol. 3, part G. Springer Verlag Berlin 1991, p. 73

[47] Nachr. Chem. Tech. Lab, 44 (1996), No. 11, p. 1080

[48] UBA Annual Report 1995, Oct. '96, p. 150.
       K. S. Booth, V. Dharmarajan (BAYER Corp.), *A. T. Jolly, J. P. Lyon* (ICI Polyurethanes), *J. L. Fosnaugh* (Dow Chem. Co.), in [60], p. 10–17, 266, 272

[49] List of limits for air (Luftgrenzwertliste) (TRGS) – 900, October 1996 Edition

[50] *R. Shaldrake:* Das Gedächtnis der Natur(The presence of the Past). Scherz Verlag, Munich, 1990

[51] *W. A. Lidy, E. Rightor, H. Phan Thanh, D. Cadolle* (Dow Chem. Co./Dow Europe S.A.) in [60], p. 119

[52] E.g. *W. Mormann* in G. *Oertel* (Ed.): Polyurethane Handbook, 2nd Edition, Hanser Publishers, Munich, Vienna, New York 1994, p. 33 (Lit. [121])
       C. D. Eisenbach in G. *Oertel* (Ed.): Polyurethane Handbook, 2nd Edition, Hanser Publishers, Munich, Vienna, New York 1994, p. 39

[53] Nachr. Chem. Tech. Lab. 45 (1997) No. 2, p. 114; original quotations from papers by G. *Maier* et al. and *A. Schulz* et al.

[54] *G. Burkhart, V. Zellmer, R. Borgogelli* (Th. Goldschmidt AG/Chem. Corp.); *J. Fis, D. W. Schumacher* (OSi Specialities S.A./Group), in [60], p. 29, 144

[55] *K. Uhlig:* Polyurethane. Kunststoffe 83 (1993) 10, p. 800

[56] BAGV Chemie e.V./VCI: "Fakten zur Chemie-Diskussion" No. 46, October 1992

[57] *H. G. Hauthal*: Renewable raw materials – perspectives for chemistry. Nachr. Chem. Tech. Lab. 44 (1996) No. 1, p. 32

[58] UT June/July 1996, p. 14

[59] *K. Uhlig*: Kunststoffe 82 (1992) 12, p. 1300

[60] *T. Heinemann, H. J. Scholl, R. Welte,* BAYER AG, in: Proceedings of the Polyurethanes EXPO '96, 20th–23rd October 1996, Las Vegas USA, p. 320

[61] Brockhaus Enzyklopädie, 19th Edition, vol. 21 (1993), p. 144

[62] Römpp Chemie Lexikon, vol. 1, 9th Edition, Thieme Verlag, Stuttgart 1989, p. 489

[63] ibid. vol. 2 (1990), p. 1212

[64] see [61], vol. 20 (1993), p. 419

[65] see [62], vol. 5 (1992), p. 4199

[66] ibid., p. 3886

[67] see [61], vol. 24 (1994), p. 604

[68] DE 19 545 550 (1994) BASF; EP 672698 (1994) Bayer
      EP 672694 (1994) Bayer
      EP 682050 (1994) Eridania Zuccherifici
      EP 259722 (1987) Henkel
      DE 3347045 (1983) Henkel

[69] UT April/May 1996, p. 22

[70] List of MAC and BAT values, 1995, VCH Verlagsgesellschaft mbH, Weinheim 1995

[71] *H. Küng*: Weltethos für Weltpolitik und Weltwirtschaft. Piper Verlag, Munich 1997, p. 215 et seq.

[72] *H. G. Hauthal*: Nachr. Chem. Tech. Lab. 44 (1996) No. 1, p. 32

[73] Bayer AG: "Desmopan guide values", published 1.12.1995/1996, order No. KU 46407

[74] Bayer AG: Quality assurance for polyurethane raw materials, order No. PU 50023, published 10/92

[75] *C. F. v. Weizsäcker:* Der Garten des Menschlichen. Hanser, Munich, Vienna 1977; published under licence in Fischer Taschenbuch Verlag, Frankfurt/M. 1983, p. 36

[76] *F. A. Henglein*: Grundriß der chemischen Technik. Verlag Chemie, Weinheim 1949, p. 215

[77] BMU press release. 39/96, 7.10.96

[78] BMU press release. 5/96, 25.1.96 and Official Federal Gazette No. 47 of 20.9.1996

[79] How rubbish is transformed into gold. Kölner Stadt-Anzeiger No. 29 of 4.2.1997

[80] *A. L. Kennedy, D. J. Aul, W. E. Braun, J. T. Hauck, E. G. Minkley* (Carnegy Mellon Univ./Res. Inst.), *R. E. Bailey* (Bayley Ass.), in [60], p. 258

[81] Technical Session B "Polyurethanes Recycling and Manufacturing Technology", in [60], p. 68–111

[82] *K. W. Dietrich, D. W. McCullough*: Method for Making Insulating Materials

from Recycled Rigid Polyurethane Foam. In: UTECH '96 Conference, Book of Papers, Rapra Technology Ltd., paper 64

[83]   UT Feb./March 1994, p. 37
[84]   ECO-PROFILES of the European plastics industry: Polyurethane Raw Materials (TDI, MDI, Polyols); Report by Prof. *Ian Boustead* for ISOPA, June 1996
[85]   UBA press release 31/95 of 22.8.1995
[86]   BMU press release 54/95 of 22.11.1995
[87]   UTECH '96, PAPER 23
[88]   UT April/May 1996, p. 14: Aprithan Schaumstoff GmbH, Abtsgmünd
[89]   UTECH '96, PAPER 22
[90]   ISOPA Recycling Polyurethanes, Fact Sheet: Options in Practice, 1997
[91]   BMU press release 46.96 of 11.11.1996
[92]   UTECH '96, PAPER 25
[93]   "direkt", newspaper for employees of Bayer AG, No. 786, November 1996, p. 3
[94]   Fact Sheets Recycling Polyurethanes: "Recovery of Rigid Polyurethane Foam from Demolition Waste", ISOPA/BING, July 1996; "Energy Recovery from Flexible PU Foam", ISOPA/EUROPUR, June 1996
[95]   BAGV Chemie e.V./VCI: "Fakten zur Chemie-Diskussion" No. 45, January 1992. Dr. Curt Haefner Verlag, Heidelberg
[96]   *W. Asenhuber:* Ökonopoly. In: Sielmanns Abenteuer Natur 6/96, p. 62
[97]   UT April/May 1996, p. 31: *Charles Churet,* Dow Europe: Keynote Presentation UTECH '96, The Hague
[98]   The Report, Foamed Plastics, No. 696; Freedonia Group Inc., USA; referred to in UT Feb./March 1996, p. 17
[99]   Economics newspaper "aktiv" of 19.6.1982

List of illustrations

Source	Figure Nos.
Hennecke Maschinenfabrik GmbH, St. Augustin-Birlinghoven	54, 55, 93, 94, 119–121
ICI: Vacpac-Panel, ICI Polyurethanes	91
ISOPA	68, 112
Krauss-Maffei, Munich	60
Leppkes, R., BASF/Elastogran: Polyurethane. Verlag moderne industrie, Landsberg/Lech 1993, p. 13	71
Neanderthal-Museum, Talstr. 300, D-40822 Mettman	1
"Neue Werkstoffe", Textbook p. 6; Fond der Chemischen Industrie in VCI, Karlstr. 21, D-60329 Frankfurt	2
Prolingheuer, E. Ch., Bayer AG, Leverkusen	87
Raßhofer, W. (Ed.): Recycling von Polyurethan-Kunststoffen. Hüthig, Heidelberg 1994, p. 19, 24, 29	112, 114, 115
Reed, D. (Ed.): Urethanes Technology, usually last page. Crain Communications, London, and ICIS-LOR, London	5
Reichmann, W., Bayer AG, Leverkusen	11
Shell Chemicals Europe	82
Steelpaint GmbH, Kitzingen	31

Index

Applications of polyurethanes are listed alphabetically in Table 1 (pages 27 to 35) and Table 2 (pages 44 to 54) and are illustrated in Figures (pages 35 to 43). They are not included in the following index apart from a few exceptions, where a specific application is used to explain the technical context.

CFCs 127–129, 142, 147, 164
Chain extender 84, 108, 121–124, 146
Chain lengths 119
Chemotronics process 91
Chipboard industry 99
Clamping units 92, 93
Coal 19, 109, 143
Coating 27, 116
Cold Box Process 99
Cold formable systems 61
Cold slabstock foams 56
Cold-cast 89
Cold-cure foam 14, 57, 74, 92
Cold-cure systems 96
Colourants 124, 134
Combustion modified high resilience
    (CMHR) foams 14, 56, 74, 159
Compression set 57–59, 64, 66, 67
Compressive strength 56–59, 61, 64
Compressive stress 57
Continuous production 89
Continuously operating plant 88
Corfam 68
Corrosion prevention 42
Counterflow injection 86
Cream time 106
Crosslinking agent 84, 96, 108, 117,
    121–124, 146
Crosslinking of PU rubber 124
Curing time 107
Cycle time 107
Cyclic esters 119
Cyclohexane-1,4-diisocyanate
    (CHDI) 112

Density 56–59, 61, 64, 65, 67, 71, 106
Dimensional stability 61, 79
DIN/EN/ISO standards 78
Diols 120
Diphenylmethane-1,4-diisocyanate
    (MDI) 140
Diphenylmethane-4,4'-diisocyanate
    (MDI) 114, 115
Discontinuous production 89
Disposal 19

Domains 138
Durable consumer goods 152

Ecological acceptability 153
Ecology 148
Elastanes 27, 28, 70, 89, 99, 120
Elongation at break 57–67, 158
Energy absorption 57
Energy conservation 37
Energy recovery 151, 154, 163
Energy-absorbing (EA) foams 59
Entropy elasticity 43
Environment 146, 152
Environmental pollution 88, 143
Environmental regulations 96
Environment-friendly paint coating 42
EP/IC 66, 74
Epoxide 102, 118, 144, 146, 161
Equivalent weights 104
Ethylene oxide 111, 118
Exothermic reaction 107
Exothermically 101
Exothermically reacting mixture 127

Fatigue behaviour 77
Fibres 22, 27, 29
Filler 124, 133, 159
Finite element method (FEM) 82
Fire regulations 73
Fire safety tests 72
Flame retardants 56, 73, 74, 124, 133
Flame-laminated foams 56
Flammability 101
Flexible foam 17, 21–23, 29, 55, 89,
    106, 114, 119, 121, 128, 130, 133, 158,
    159, 164
Flexible foam flakes 159
Flexible foam slabstock 14, 94, 95
Flexible moulded foams 24, 57, 58, 74,
    92, 116
Flexible padding foams 93
Flexible slabstock 23, 140
Flexible slabstock foams 55, 74, 75, 90
Flexible structural foam 17, 22, 25